VALUE MANAGEMENT IN DESIGN AND CONSTRUCTION

Also available from E & FN Spon

Facilities Management
K Alexander

The Construction Net
A Bridges

Understanding JCT Standard Building Contracts
5th edition
D M Chappell

Creating the Built Environment
L Holes

An Introduction to Building Procurement Systems
J W E Masterman

Construction Contracts
2nd Edition
J Murdoch and W Hughes

Building International Construction Alliances
R Pietroforte

Understanding the Building Regulations
S Polley

Risk Analysis in Project Management
J Raferty

Programme Management Demystified
G Reiss

Project Management Demystified
2nd Edition
G Reiss

Risk Avoidance for the Building Team
B Sawczuk

The Building Regulations Explained
J Stephenson

For more information about these and other titles please contact:
The Marketing Department, E & FN Spon, 11 New Fetter Lane, London, EC4P 4EE.
Tel: 0171 842 2180

VALUE MANAGEMENT IN DESIGN AND CONSTRUCTION

The economic management of projects

JOHN KELLY
Department of Building Engineering and Surveying,
Heriot-Watt University
Edinburgh, UK

and

STEVEN MALE
Department of Civil Engineering,
University of Leeds
Leeds, UK

Taylor & Francis
Taylor & Francis Group
LONDON AND NEW YORK

First published 1993 by Taylor & Francis, an imprint of Taylor & Francis
Reprinted 1994

Reprinted 1998 by Taylor & Francis, an imprint of Routledge
2 Park Square, Milton Park, Abingdon, Oxon, OX14 4RN
270 Madison Ave, New York NY 10016

Transferred to Digital Printing 2005

© 1993 John Kelly and Steven Male

Typeset in 10/12pt Garamond by Expo Holdings, Malaysia

British Library Cataloguing in Publication Data
A catalogue record for this book is available from the British Library

Library of Congress Cataloging in Publication Data
A catalog record for this book is available from the Library of Congress

ISBN 0–419–15120–6 (pbk)

DEDICATION

This book is dedicated to Bob Charette, who produced one of the most gruelling interview schedules we have ever encountered, and Howard Ellegant, whose insights into functional analysis have left us unsure of why a bus stop exists.

CONTENTS

PREFACE

Value management is a service which maximizes the functional value of a project by managing its evolution and development from concept to completion, through the comparison and audit of all decisions against a value system determined by the client or customer.

Value management is an emerging service in Europe and this book represents the authors' current thinking on what is a continually evolving topic. The term 'value management' is chosen for use here rather than the more popular North American term 'value engineering', differentiating between a broad management approach to value rather than a narrower focus on technical performance. The latter is viewed by the authors as only one component of the management of value. Other components in the 'value equation' are the interactions between time, cost and quality in the context of the client's strategic management process and subsequent value system. Value management has also now become an accepted term within Europe, and particularly at European Community Directorate level.

The North American framework of value engineering has been used in this book as an example of a widely used model of the service in order to gain insights into the use of the philosophy, tools, techniques and procedures of value management in construction. The authors recognize explicitly, and have argued consistently in other publications, that the straight implantation of this service from a different construction culture is inappropriate to Europe. One immediate distinguishing feature here is the existence in the UK of a professional cost consultant as a member of the design team. In North America the cost consultant is a technical advisor employed by the architect or engineer.

In order to explore and discuss the emerging service of value management in a UK context, the authors have distinguished between services that they have defined as:

- Value management
- Cost management

One of the major objectives of the book is in comparing and contrasting these services. The authors have also defined the term 'project economics' as encompassing both of these services within a broader framework for the economic management of projects. A proposed UK methodology for project economics is outlined in Chapter 12.

There is evidence to suggest that a non function-oriented North American style of value engineering service is being offered in the UK. The authors are of the view that this service does not maximize the potential of the UK construction culture. There is therefore a danger already of an adverse reaction from UK clients and designers to value engineering.

This book represents the culmination of a number of years of research and the authors would like to thank Madeleine Metcalfe at E & F N Spon for her considerable patience and support during its development. We would also wish to thank all those people that we have interviewed in Canada, the US, mainland Europe and the UK who are too numerous to name here.

We would also like to gratefully acknowledge the financial support of the Royal Institution of Chartered Surveyors Education Trust and the guidance and encouragement of the RICS Quantity Surveyors Division Research Group throughout this research.

John Kelly
Steven Male

Heriot-Watt University
Edinburgh

PART ONE

THE DEVELOPMENT AND PRINCIPLES OF NORTH AMERICAN VALUE ENGINEERING

CHAPTER 1 _____

INTRODUCTION TO VALUE MANAGEMENT

_____

1.1 A DEFINITION OF VALUE MANAGEMENT

Value management is the name given to a service in which the sponsor of a project, the client, transmits a clear statement of the value requirements of that project to the project designers. Once realized, the client's value system can be used to audit:

- The client's use of a facility in relation to its corporate strategy;
- The project brief;
- The emerging design;
- The production method.

The characteristics which separate value management from an accounting vision of an audit are:

- A positive and pro-active approach through the use of a multi-disciplinary team-oriented creative process to generate alternatives to the existing solution;
- The use of a structured systems method;
- The relationship of function with value.

Value management should not be seen as a conflict-oriented design review, cost reduction or standardization exercise.

Maximum value as defined by Burt (1975) is obtained from a required level of quality as least cost, the highest level of quality for a given cost or from an optimum compromise between the two. Value management is therefore the management of a process to obtain maximum value on a scale determined by the client. A number of techniques are identified as being useful in this enterprise but none are compulsory.

A useful one sentence definition of value management is:

> a service which maximizes the functional value of a project by
> managing its development from concept to completion and
> commissioning through the audit (examination) of all decisions
> against a value system determined by the client.

1.2   BACKGROUND TO VALUE MANAGEMENT

A number of authors, Markus (1967), Broadbent (1973), Jones (1981) and
Cross (1989) have discussed design method in the context of building
design. They have investigated both formalized and philosophical
approaches to the provision of maximum value. However, in North
America, coincidentally over the same period of time, a technique originally developed for manufacturing was being refined and applied to
construction. This North American technique is marketed under the name
'value engineering'.

This book describes and reviews North American value engineering as it
is represented by both books and practice. The better parts from this
review are integrated with established design method and management
techniques into a system for full value management in a British context.
The authors have adopted the term 'value management' rather than the
term 'value engineering' since the former is widely used throughout
Europe. It has also been adopted by the European Community in its
SPRINT programme (Strategic Programme for Innovation and Technology
Transfer) as an important business procedure. Furthermore, the term value
management encapsulates a broader impetus for the management of
'value' for the client, commencing deep within the client organization and
related to its on-going strategic management process, of which the
contracted-out design and construction services provided by the industry
are only one contributing part.

1.3   VALUE MANAGEMENT AND THE CHANGING BRITISH
CONSTRUCTION SCENE

This section briefly reviews the changing construction scene in the UK in
order to set value management within this context. A number of forces
have been at work in the past decade that have resulted in major changes
in the way that the construction industry now operates and will operate in
the future. These forces can best be summarized as:

• A stable government committed to the philosophy and ideology of
   the 'market';

- A macro-economic climate that has changed from recession to boom to recession over the decade;
- The introduction of competitive fee bidding for construction consultants' services;
- The emergence of a diversity of procurement routes for projects;
- Clients that have become more demanding and knowledgeable of the construction process;
- The emergence of the European dimension with the establishment of the Single European Market from 1993 onwards;
- A re-definition of roles in the industry;
- A move towards single-point responsibility and management of the total construction process.

A re-definition of roles in the industry

A number of major reports have been issued within the last ten years that have suggested the redefinition of roles that are now occurring in the industry (MAC, 1985; NEDO, 1988; PRS, 1987; RICS, 1991). Also, there is no doubt that competitive fee bidding has accelerated the rate of change when coupled with the other forces at play in the industry. There are three major groupings that are now changing the manner in which they are responding within the industry, namely, construction firms, the surveying and design professions.

The large construction companies are diversifying their service base and involvement throughout the construction process. They have moved increasingly towards offering services directly for the client and have also become involved in property development. They are using their management skills to manage 'the process'. The surveyors in private practice, especially the large practices, are diversifying their service base away from traditional roles towards managing the total process. There is an increasing number of surveyors now working in client organizations for 'knowledgeable clients', who are also able to have a greater input into the total management of 'the process'. Finally the designers, the architects and engineering consultants, are increasingly finding their management/ supervisory roles under threat, primarily from the surveyors and construction firms, and this has been exacerbated by the alternative procurement options which are now allowing different administrative and contractual choices to be exercised for managing design and construction. In addition, the traditional roles of the architectural and engineering consultants have stressed a design orientation and this is now being procured as the primary service either as a function of client **or** consultant choice. Furthermore, in a complimentary fashion to the surveyors, the design professions are also working in greater numbers in-house for client organizations and assisting them in managing 'the process'.

This re-definition of roles, primarily in the larger organizations within construction, has resulted in a convergence of different types of organization towards offering a service to manage the total process for the client in order to gain a competitive edge in the market place. This has resulted in the notion of providing 'added value' in service provision coupled with a diversification of commercial risk away from a narrowly defined technical core business. The result will be increasing competition among larger firms – the construction companies and consultancies – to manage the total process and a restructuring of the industry that will create pressures on the medium and small sized technically based firms.

This re-definition of roles is also creating the requirement for a new skill base within the industry for managing the total process.

The new skill base

The move towards managing the total process for the client either in-house or by the larger organizations in the industry requires two distinct but interrelated sets of skills – skills of 'strategy' and skills of 'implementation'. This skill base requires an ability for: conceptual thinking in, and problem solving through the use of, teams of specialists.

Skills of implementation involve: people skills; organizational skills; programming skills; negotiations skills; resource allocation skills and team skills.

These two distinct sets of abilities come together under the generic title of skills for 'managing projects'.

The framework for managing projects

The perspective adopted in this book places value management in the context of contributing towards the management of the total process, that is, the management of the construction process as a project from its inception to completion and commissioning. The framework for managing the total process is effected by three interlocking components:

1. The organizational framework under which the construction process is administered;
2. the procurement strategy that is adopted that sets up the organizational and administrative relationships between the constituent parties;
3. the legal framework that binds the parties together contractually.

In this context the management of projects is concerned with controlling: time; cost; quality. In the context of function, which is client determined and assessed through the tools and techniques of value management.

1.4   AN OUTLINE OF THE BOOK

The book is in four parts. Part One gives a background to the development of value engineering in North America and describes the principles evolved during the first twenty years of its use.

Part Two describes in detail the use of value engineering in North American construction. Chapter 3 provides an account of North American value engineering practice applied to construction. Chapter 4 outlines four diverse case studies as examples of value engineering in practice. Chapter 5 outlines three different implementation strategies adopted by North American clients, two of which could be described as being successful and the third less so. Chapter 6 is a case study of one of the major value engineering consultancies in the States.

Part Three is a critique and analysis of the implementation of value engineering in North America and abstracts the principles learned from this research work for further elaboration in Part Four. Chapter 7 is a critique of North American value engineering based on data gathered through case studies, interviews and questionnaires. Chapter 8 distinguishes between cost and value management in a European context and introduces the notion of project economics and the client value system related to project management.

Part Four develops the insights gained from the earlier chapters into a proposal for UK implementation of value management. Chapter 9 explores in detail functional analysis, the cornerstone of value management practice. Chapter 10 describes the principles of life-cycle costing as a decision making tool in value management. Chapter 11 discusses group dynamics and team skills in the context of the leadership of value management studies, providing a number of tools and techniques that can be utilized in practice. Chapter 12 introduces a proposed UK value management methodology. Chapter 13 identifies key issues to be considered by practice in the application of value management.

CHAPTER 2 _____

NORTH AMERICAN VALUE ENGINEERING PRINCIPLES

_____

2.1 BACKGROUND AND DEFINITION

North American value engineering provides a well-documented basis upon which to build a value management method and is therefore considered in this context.

Value engineering is based on the work of Lawrence Miles who, in the 1940s was a purchase engineer with the General Electric Company (GEC). At that time, manufacturing industry in the United States was running at maximum capacity which resulted in shortages of a number of key raw materials and components. GEC wished to expand its production and Miles was assigned the task of purchasing the materials to permit this. Often he was unable to obtain the specific material or component specified by the designer so Miles reasoned, 'if I cannot obtain the product I must obtain an alternative which performs the same function'. A characteristic of value engineering from the beginning was the team approach to creativity which allowed the generation of many alternatives to the existing solution. Where alternatives were found they were tested and approved by the product designer.

Miles found that many of the substitutes were providing equal or better performance at a lower cost and based on these observations he proposed a system which he called **value analysis**. The definition of value analysis is:

> an organized approach to providing the necessary functions at the lowest cost

From the beginning, value analysis was seen to be a cost validation exercise which did not affect the quality of the product. The straight omission of an enhancement or finish would not be considered value analysis. This led to the second definition:

> Value analysis is an organized approach to the identification and elimination of unnecessary cost

Unnecessary cost is:

> Cost which provides neither use, nor life, nor quality, nor appearance, nor customer features

In this context:

1. Use refers to the utility of the component. This utility is measured by reference to the extent to which it fulfils the required function.
2. The life of the component or material must be in balance with the life of the assembly into which it is incorporated. For example, unnecessary cost may be incorporated if a component is specified which has a life of 12 years within a product which will be redundant in four years.
3. Quality is a subjective function, but however it is perceived it must be preserved. The philosophy of value engineering looks towards reducing cost without sacrificing quality.
4. The appearance of a product is often one of the most important features to a customer. In appraising a motor car the appearance of the outside and passenger compartment will not be questioned.
5. Customer features are those which sell. The graphic often applied to products is an example.

2.2   US GOVERNMENT PATRONAGE

In 1954 the US Department of Defense Bureau of Ships became the first US Government organization to implement a formal programme of value analysis. It was at this time that the term value engineering came into being for the administrative reason that engineers were considered the personnel most appropriate for this programme. The term value engineering came into common use and will be used throughout this book to denote the type of practice used in North America.

Table 2.1   VE activity by key US government agencies

1963	Introduced into US Department of Defense, Navy Facilities Engineering Command.
1965	Introduced into US Department of Defense, Army Corps of Engineers.
1968	Used by Facilities Division of the National Aeronautics and Space Agency.
1973	US General Services Administration, Public Building Service published the first VE service contract clauses for use in Design and Construction Management contracts.
1974	US General Accounting Office publish (May 6, B-163762) 'Need for increased use of value engineering, a proven cost saving technique, in federal construction'.
1976	US Environmental Protection Agency, Water Program Operations, publish a 'VE Workbook for construction grant projects' MCD-29, making VE mandatory on projects over $10 million.

The formation of the Society of American Value Engineers in 1959 established the technique and the name. Value engineering spread to many US federal, state and local government agencies following the cost reduction programme of Secretary McNamara in 1964. Within construction, government agency action, as illustrated in Table 2.1, was instrumental in the furtherance of the technique.

Although US private sector industry warmed to the technique it was never accepted with the same enthusiasm as that of the Government agencies. The reasons for this will be discussed in Chapter 4.

2.3 THE JOB PLAN

A characteristic of North American value engineering is the team approach to creativity through application of the job plan, a logical, sequential, approach to the study of value. All North American value engineering texts refer to a pattern derived from Miles' original work which is summarized in Table 2.2. It is to be noted that Miles was solely concerned

Table 2.2 Miles' original job plan 1961

Phase 1: **Orientation**
What is to be accomplished, what does the client need and/or want, what are the desirable characteristics?

Phase 2: **Information**
Secure all costs, quantities, drawings, specifications, manufacturing methods, samples and prototypes. Understand the manufacturing process. Determine the amount of effort which should reasonably be expended on the study.

Phase 3: **Speculation**
Generate every possible solution to the identified problem using brainstorming sessions. Record all suggestions.

Phase 4: **Analysis**
Estimate the dollar value of each idea and rank in order of highest gain and highest likely acceptability. Investigate thoroughly the best ideas.

Phase 5: **Programme Planning**
Establish the manufacturing programme by identifying operations, design and production personnel, suppliers, etc. Promote an ethos of creativity in all involved parties.

Phase 6: **Programme Execution**
Pursue the programme, evaluating and appraising further suggestions from suppliers, etc.

Phase 7: **Status Summary and Conclusion**
If in a position to take executive decisions then act on new ideas, if not make recommendations to those who are to make the decision.

with manufacturing and this is reflected in his job plan. Those subsequent authors whose concerns included construction modified the job plan to accord with the processes and terminology of the industry. The review which follows contains the essential ingredients of the North American job plan applied to construction.

Phase 1: Orientation

The orientation meeting, promoted by the General Services Agency and practised by, *inter alia*, New York City Office of Management and Budget, is held following the appointment of the design team.

It is a meeting chaired by the value engineer for the project and attended by the design team and those client representatives who have an interest or possess some ownership of the problem being addressed. The objective of the meeting is to pose the questions asked by Miles namely:

- What is to be accomplished?
- What does the client need and/or want?
- What are the desirable characteristics?

As practised by New York City, it is an opportunity to allow everyone involved in the project to understand all the issues and constraints. It provides everyone who is to make a decision an opportunity to give and receive information. The value engineering procedure of the New York City Office of Management and Budget is reviewed in Chapter 3.

Phase 2: Information

In this phase all of the available information relating to the project under review is gathered together. The objective of the information gathering is to identify the functions of the whole or parts of the project, as seen by the client organization, in clear unambiguous terms. The information should not be based upon assumption but be obtained from the best possible source and corroborated, if possible, with tangible evidence. The reasoning behind this is that the quality of decision making cannot rise above the quality of the information upon which the decision is to be made.

The specific information being sought at this stage is:

(a) Client needs are the fundamental requirements which the project must possess to serve the client's basic intentions. Needs should not be seen solely in terms of utility as the client may have a need for a flamboyant statement or a need for a facility which heightens the client's esteem.

(b) Client wants are the embellishments which it would be nice to have but do not satisfy need.

(c) Project constraints are those factors which will impose a discipline upon the design. For example, the shape of the site, planning requirements, regulations, etc.

(d) Budgetary limits expressed as the total amount which may be committed to the project in initial capital and life-cycle costing terms.

(e) Time for design and construction as well as the anticipated period for which the client will have an interest in the building. This latter concept is discussed in more detail in Chapters 9 and 10.

Although, as stated above, the quality of decision making cannot rise above the quality of the information upon which the decision is to be made, care should be taken not to spend unjustifiable time and effort in information seeking. Fallon (1971) refers to the dilemma between the dangerous consequences of acting upon inadequate information and the possible missed opportunity when waiting for reliable information to arrive.

As an aid to information gathering it is considered by some authors that the value engineering team should construct a Functional Analysis System Technique (FAST) diagram. This function-logic diagram, resembling a decision tree, is commenced with the prime function to the left and constructed to answer the questions 'why?' when reading from right to left and 'how?' when reading from left to right. It is argued that the very action of drawing the FAST diagram concentrates the minds of the value engineering team on the functional requirement of the project.

The concept of function and the FAST diagram is considered in detail in Chapter 7.

Whether or not a FAST diagram is used, a prime task in the information stage is the realization of those items in the current brief or design which attract high cost for low-functional utility and/or those items which have a high importance but attract a low cost.

Although not a feature of Miles' first edition, North American practice has evolved a tradition that functions are recognized in a simple verb plus noun form. A wall for example may have the functions: support roof, protect from weather, maintain internal temperature etc. It is these functions that the value engineering team will answer in terms of technical solutions.

Phase 3: Creativity

In the creative phase the value engineering team puts forward suggestions to answer the functions which have been selected for study. It will be noted from the case studies in later chapters that only a few cost-dominant functions will normally be selected for study.

There are a number of creative techniques, for example; brainstorming, the Gordon technique, the synectics technique and many more. Creativity as a group activity is discussed in Chapter 11, however, for illustrative purposes only, brainstorming is described in outline below.

Brainstorming is the most popular technique used by value engineers to generate ideas in the creativity phase. The technique requires that a group consider a function and contribute any suggestion which will answer that function. Every suggestion, no matter how apparently stupid, is recorded. So, for example, in the 'wall' illustration given above, suggestions for a butchers' cold store for the function 'maintain internal temperature' could be: ice, cold air curtain, insulation, vary pressure, in fact, any idea that comes to mind. Obviously a large number of the ideas will be disregarded because a further function of the wall is to 'support roof'.

There are various rules which apply to the management of a brainstorming session of which the two most important are: firstly, no criticism of any suggestion by word, tone of voice, gesture or any other method of indicating rejection is allowed. Secondly, the exercise is one of generating as many suggestions as possible. The good suggestions will be randomly scattered amongst all suggestions. Research has indicated that in any sample, the number of good suggestions remains fairly constant as a proportion of wild suggestions, so the more suggestions that there are, the more good suggestions will be obtained. All suggestions are recorded and none are rejected on the grounds of apparent irrelevance.

Research has also shown that original suggestions are as likely to come from those inexpert in a subject as those who are expert. In other words the interior design consultant may come up with a good original suggestion for the solution to a structural function. One reason put forward for this is that the consultant will not be constrained by professionally-determined technical rules or education.

Phase 4: Evaluation

The value engineering team evaluate the ideas generated in the creativity phase using one of a number of techniques many of which depend upon some form of weighted vote. This stage forms a crude filter for reducing the ideas generated to a manageable number for further study.

Phase 5: Development

The accepted ideas, selected during phase 4, are investigated in considerable detail for their technical feasibility and economic viability. Outline designs will be worked-out and costs realized. There is wide scope for the use of life-cycle cost models and computer-aided calculations at this stage, indeed it is conceivable that this stage could not be undertaken in a reasonable time without the aid of computer facilities.

At the end of the development stage the team will again consider the worked-up ideas. Those which do not comply with the basic value engineering philosophy will be dismissed. That is, all ideas which either cost more than the original or are found to reduce quality are rejected.

Phase 6: Presentation

The refined ideas supported by drawings, calculations and costs are presented by the value engineering team to the body which commissioned the value engineering exercise.

Phase 7: Feedback

It is important that the value engineer receives some detail of those ideas that have been put into practice and be given the opportunity of testing the design and cost predictions of the team.

2.4   FUNCTION, VALUE, COST AND WORTH

These four terms are defined within the context of North American value engineering as follows:

(a) Function is a characteristic activity or action for which a thing is specifically fitted, used or for which something exists. Therefore something can be termed functional when it is designed primarily in accordance with the requirements of use rather than primarily in accordance with fashion, taste or even rules or regulations. Value engineers distinguish between a basic function and a secondary function.

A basic function is defined as the performance characteristics which must be attained by the technical solution chosen. Secondary functions are the performance characteristics of the technical solution chosen other than the required basic function. For example, a basic function of a window may be to transmit light. A technical solution to the function could be a sheet of glass. Secondary functions of the required design because of the choice of a sheet of glass as the solution are to prevent solar glare, control heat gain, prevent condensation, and prevent cold radiation.

(b) Value is a measure expressed in currency, effort, exchange or on a comparative scale, which reflects the desire to obtain or retain an item, service or ideal. Authors have prefixed value with the terms use, exchange, esteem, aesthetic, judicial, moral and religious in order to highlight different value situations. Use value, for example, is a measure of the function of the item, for example a chisel has high value as a tool for cutting wood but low value as a screwdriver although it could be used for both tasks. Exchange value is the amount for which an item may be sold. Esteem value is the amount an owner or user is prepared to pay for prestige or appearance.

(c) Cost is the price paid or to be paid. It is often said that one man's price is another man's cost. Cost can be divided and distributed among elements and to some extent, functions.

(d) Worth is defined by North American value engineers as the least cost to perform the required function or the cost of the least cost functional equivalent.

The definitions above are a fair summary of the views of authors of North American value engineering texts in that none hold an opposing view. However, there is considerable room for discussion particularly relating to the extent to which functions can be priced. It may be the case that functions can be appreciated in terms of cost but not actually costed. For example, the cost of a window that transmits light through the medium of a sheet of glass may be costed. The additional cooling load required to control heat gain through the glass may also be costed. The alleviation of the cold-radiation feeling of someone whose desk is next to the window is less easy to cost. However, what can not be disputed is the advantage of being able to see the relative costs of the prime function and the secondary functions.

Worth is another definition which is debatable. As defined, a high-quality door in a bank is worth the cost of a hardboard-faced hollow door with the cheapest ironmongery. This is not a useful definition unless it is appreciated that the use value of the door is the cost of the basic solution and the difference in the cost of the two alternatives represents esteem value.

2.5   THE TIMING OF THE APPLICATION OF VALUE ENGINEERING

Value engineering is a philosophy rather than an absolute method or set of rules. The philosophy of sitting back and considering a design decision can be carried out, therefore, by any specification writer or designer at any stage in the design. However, during the past forty years, value engineering has attracted techniques and methods which permit exercises to be carried out as formal review procedures.

The procedures tend to be associated with stages in the design and construction process and, therefore, a particular procedure will be used at a particular stage. A common feature of all procedures is that they all consider the cost to the client as being the total cost during the period that the client maintains an interest in the building. all procedures therefore depend upon the use of life-cycle costing (cost in use) techniques.

All value engineering authors agree that the maximum cost reduction potential occurs early in the briefing/design process. For example, an analysis of the brief for a hospital project may find that tests undertaken in two 30 m² laboratories are not incompatible and may be undertaken in one 45 m² laboratory with a reduction in technician support. This represents a considerable life-cycle cost saving. If this fact is recognized at the briefing stage the cost of implementation is low. If the opportunity for saving is recognized at the design stage, when the form of the building is complete and detailed design has been commenced, the alteration is not impossible even if the two laboratories are remote from each other in the design. However, alterations in design are expensive in consultant time and in terms of the disruption to the design programme. This concept is illustrated below.

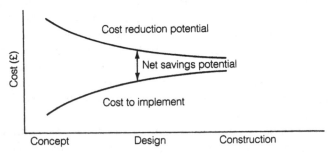

Figure 2.1   A simple representation of the opportunity to change design.

A report by the US General Accounting Office (GAO, 1978) confirmed this view and put forward the suggestion that more structured effort should be made at the early design stage when the project was incurring the lowest cost. GAO were of the firm belief that while expenditure during design was comparatively small, the decisions and resource commitments made during this phase significantly influenced later resourcing and expenditure by greater orders of magnitude. This is illustrated in Figure 2.2

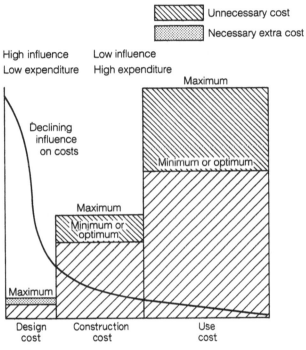

Figure 2.2 A small increase in design costs increase the chance that construction and use costs are at their minimum.

The points in time at which value engineering is commonly carried out are at the end of the formulation of the brief (10% design); at just prior to working drawings (35% design); and during the construction process.

2.6   SUMMARY

All North American texts on the subject of value engineering refer to, develop and discuss the job plan. Its principles are judged to be sound;

indeed it resembles in many respects the classical *modus operandi* of research. It will be used in this text as a foundation to a recommended value management methodology and therefore will be frequently revisited.

The next chapter deals with value engineering practice, outlining the diversity of approaches encountered, the procedures which may be used at various project target points, all of which are built around the phases of the job plan.

PART TWO

THE PRACTICE OF NORTH AMERICAN VALUE ENGINEERING

CHAPTER 3 ────────────────

NORTH AMERICAN VALUE ENGINEERING PRACTICE

─────────────────────────────

3.1   INTRODUCTION

North American textbooks tend to portray a standard approach to the application of value engineering. However, during research in USA and Canada it was found that while the job plan was followed there were various application procedures at different stages of the project. There are four clearly identifiable formal approaches to value engineering and a number of variations on the same themes.

The four formal approaches are defined as:

(a) *The charette*, The meeting following the compilation of the client's brief, attended by the full design team and by those members of the client's organization who have contributed to the brief. This meeting is conducted under the chairmanship of the value engineer;

(b) *The 40 hour study*. This is an examination of the design developed to sketch-design stage. This is carried out by an independent team of design professionals, who have not been involved with the design until the time of the study, again under the chairmanship of the value engineer;

(c) *The value engineering audit*. This is a study of the proposals made by a subsidiary of a large holding company for a vote of capital to fund a project. This study will be undertaken by a value engineering team in order to ensure that the parent company is receiving value for money;

(d) *The contractor's change proposal*. In this situation a clause in the construction contract allows the contractor to suggest changes to the proposed design in order to reduce construction costs. The contractor receives a bonus in exchange for the proposal.

These four approaches and the variations upon them are discussed in more detail below. In considering the various approaches the differences between construction cultures should be borne in mind as these are important if a value management method is to be introduced in Europe.

3.2 NORTH AMERICAN CONSTRUCTION CULTURE

In this brief analysis reference will be made back to the traditional culture of UK style construction in which the architect is the design team leader with responsibility for taking the brief and the cost limit from the client, preparing sketched alternatives, discussing the financial consequences with the chartered surveyor and engineering matters with the various engineering consultants. The chartered surveyor is responsible for preparing the cost plan which should equate with the cost limit, financially controlling the evolving design by calculating the cost of the design as it is developed by the architect and the various consultants and comparing this with the cost plan. Any danger of overspending results in referring back to the design consultants for an alternative design solution. The chartered surveyor also prepares the tender documentation which will contain a detailed description of the works but often not a full set of working drawings.

In North America there is no chartered surveyor. The engineer enjoys a much higher status than in the UK and may be the design team leader in place of the architect. The architect/engineer has responsibility for taking the brief and the cost limit from the client and is responsible for preparing a cost plan. Often this cost plan is sub-contracted to a cost consultancy. The cost consultancy is therefore employed by the architect to carry out an estimate of the tender value based upon the sketch drawings. There is no financial control service. The cost consultancy enjoys very little status and employ in the main personnel who have trained as contractors' estimators.

In the development of the sketch scheme the architect and engineers work in parallel, such that at sketch-design the architects' and engineers' draft schemes are available. This is in stark contrast to the UK where the decision to introduce a prime cost sum into the tender documentation for mechanical and electrical works might mean that these engineering installations are designed after the architect's scheme is complete and the contractor's tender has been accepted.

North American tender documentation comprises drawings and specifications. Often the specifications include performance specifications for work to be designed by the contractor.

The North American system tends to be very linear with little scope for change, in contrast to the UK style approach which iterates and has the appearance of being designed for change.

3.3   THE FORMAL APPROACHES TO NORTH AMERICAN VALUE ENGINEERING

The Charette

This method seeks to rationalize the client's brief through the identi-fication of the function of key elements and the spaces specified. This analysis through function at a meeting, involving the client's staff and the design team, should ensure that the latter understand fully the require-ments of the former. During the research into this subject interviews were held with a client who had commissioned a value engineering study following a tender in which the lowest was 18% higher than the permitted budget. The client stated that a value engineering study at briefing would be useful:

> '..just to focus on what are the objectives, what are the primary functions of any given activity and this is one area of value engineering where it was clear when we got around the table after the design was done, including the players who were involved in the design, that what each of them thought was the primary function was different and this was after the design was done.'

It should be noted that this particular study revisited the brief following the receipt of the tender. Although grossly inefficient in terms of the time spent on re-design it did enable the project to become viable. A true charette will study the brief following its compilation.

There is a theory which states that the brief given by the client to the design team is an amalgam of the 'wish lists' of all the parties who con-tribute to the brief. This is particularly so for buildings which are to house organizations comprised of diverse departments such as hospitals, univer-sities, prisons, owner-occupied offices for complex organizations and manufacturing plants. Even where a project manager is employed there is a high probability that the brief will reflect data gathered from depart-mental heads who will seek to maximize their requirements. In a prison, for instance, two departments may each have a requirement for a therapy area and the two areas may be identical, but this is not likely to be re-alized unless a study is made of the function of each space.

The charette is organized along the traditional job plan lines, the first stage being to gather as much information as is available regarding the function of the spaces defined in the brief. The function of all of the spaces are defined along with performance criteria eg., this space must be held at a constant 20°C ± 5° where the activity within the space generates heat.

The next stage in the process is to be creative in terms of arrangement, adjacency, timetabling, etc. It may be found for example that by siting two particular hospital departments together that they may use the same laboratory with immense savings in capital and running costs (including laboratory staff).

The ideas generated are recorded and analyzed and the final decisions are incorporated into the brief. The charette has five major advantages.

Firstly, it is considered by many clients to be inexpensive. There is, of course, staff time to consider and the time of the design team. However, if the design team were informed at the time of their selection of this meeting, it is suggested that their fee bid would not be greatly affected. The only real expenditure is the fee of the value engineer himself. One client interviewed during this research also stated that it was beneficial to have a cost consultant at this meeting. There is finally the secretary to the team who will either be provided by the client or by the value engineer.

Secondly, the exercise was considered by some clients to be the best way of briefing the whole team. One industrial client with a large and expanding building stock stated that even if the exercise did not realize any great rationalization the very fact that all members of the team: the architect, structural engineer, mechanical and electrical engineers, etc., were present meant that all understood fully their requirements.

Thirdly, the exercise occurs early in the process, stated by many to be the most cost-consequent stage.

Fourthly, the exercise can be carried out in a short period of time, only the most complex projects will involve more than two days work.

Finally, the exercise cuts across organizational, political and professional boundaries. One central government organization client stated that a meeting of this kind would not normally be possible since departmental heads would be reluctant to give up the time. Also, if held the meeting would be politically structured. The design team themselves would not normally be sufficiently pro-active in the organization of such a meeting. The fact that it was a value engineering exercise under the chairmanship of an outsider made it happen.

The charette is therefore an inexpensive means of examining the client's requirements by the use of functional analysis and allowing rationalization and full design team briefing.

The 40 hour value engineering study

The 40 hour study is the most widely accepted formal approach to value engineering, indeed the initial training of value engineers as laid down by the Society of American Value Engineers (SAVE) is based on a 40 hour training workshop. The study involves the review of the sketch design of project by a second design team under the chairmanship of a value engineer. It applies to all of the stages of the job plan within a working week and is seen as being quick and effective. The procedure for the study is as follows:

The client should inform the members of the design team at the time of their fee bid that the project will be the subject of a value engineering exercise. This is important both from a human relations aspect and also from the point of view of establishing how the design team are to cover the cost of any re-design work arising out of the exercise. Some clients require the members to cover this cost within their fee bid. Others state at the time of the fee bid that the design team members will be reimbursed for any necessary re-design work on an hourly basis.

The client appoints the Value Engineering Team Co-ordinator (VETC). The value engineer, in discussion with the design team, establishes the date for the study. Normally the VETC will submit a fee bid which covers the cost of the complete value engineering exercise described below.

The VETC will appoint a value engineering team, normally six to eight professionals in a mix that reflects the characteristics of the project under review. So, for instance, a project with a large amount of mechanical and electrical servicing may attract a team including four members with these professional backgrounds. These team members will be drawn from professional practice and may or may not have any previous value engineering experience. The team members are paid by the VETC.

The study is normally held near the site of the proposed project, either in a hotel or in a room provided by the client within the client's office.

The date of the study is a key date for the design team and the value engineering team. The design team must be complete to sketch-design stage one week before the date of the study. This includes the architectural design and also the structural, mechanical and electrical engineering designs. The completed drawings are sent to the VETC for distribution to the team during the week preceding the study.

During the week of the study the team will follow strictly the stages of the job plan. It is the logical step-by-step approach to the generation of alternative technical solutions which makes value engineering unique.

(a)   Monday – phase 1 (information)

The team have each had the project sketch drawings, initial cost estimate together with calculations and outline proposals for the structure and services for two days and will have gleaned some information from these. At the beginning of the study the architect and the client are present. The

VETC gives an introduction and states the objectives for the week. Often the VETC will have prepared a timetable and may also have prepared an elemental breakdown of the initial estimate.

Following the introduction the client and the design architect present the project, answer questions and the client reaffirms which areas of the project are within the scope of the exercise. This latter point is important since, for example, if the client has already reached an agreement with a trade union that a specific number of men will be employed within a plant then all ingenuity on the part of the value engineering team to reduce manning levels will be in vain. After the presentation the client and architect leave.

The team now concentrate their efforts on identifying the functions of the various parts of the building. In the study emphasis is given to those functions which are not important, or are secondary, but attract a high cost. Attention will also be paid to those functions which are primary and important but attract a low cost.

In one study for the modernization of a boiler house on a large military site in North America, with an estimated project cost of $71 500 948, the team identified 17 functions of which seven were selected for study.

(b)   Tuesday morning – phase 2 (creativity)

During this phase the group brainstorm ideas to satisfy the identified functions. In the boiler house example above over 200 ideas were generated during this session. Creativity is a rapid but exhausting process, 200 ideas could easily be produced in one hour.

(c)   Tuesday afternoon – phase 3 (judgement)

At this stage the team decide which of the ideas generated are worthy of further development. For example, of the 200 plus ideas generated above only 42 were thought good enough for development.

Before moving on to the development phase some value engineers prefer to invite the design architect back to the study at this point to discuss the acceptability of the ideas in principle. This can reduce abortive work if, for instance, the design team had already thought of the idea and rejected it or if the architect would not agree to such an idea under any circumstances.

(d)   Wednesday and Thursday – phase 4 (development)

During this phase the team may split into individual or small groups to work on the ideas in detail. The aim is to develop the ideas into a worked and costed solution.

(e)   Friday – phase 5 (recommendation)

In the morning of the final day the group reconvene to discuss the
worked solutions. At this stage those solutions which either cost more
than the original, reduce quality or are not technically feasible are
rejected. In the boiler house case above 15 worked solutions were
rejected at this stage leaving 27 viable solutions for presentation to the
client and design architect.

In the afternoon those worked solutions accepted by the team are presented to the design architect and the client.

The formal study is now at an end. The members of the study return to
their practice leaving the VETC to take away the week's work and write
the report.

(f)   The following week – action and feedback

In the early part of the following week the completed report is sent by
the VETC to the client and design architect. At this stage, one North
American government department takes all of the ideas, sets them out
on a sheet horizontally with vertical columns for each member of the
design team who receives a copy. The team members are requested to
enter either 'accepted', 'rejected', or 'further discussion required' against
the suggestions. A meeting is called where all members of the design
team gather to discuss the suggestions. All those which have been
unanimously accepted are required to be incorporated into the design.
In respect of the others discussion takes place to determine which may
be acceptable. The client will wish to be convinced of the need for
rejection.

In the boiler house example above, 11 of the 27 suggestions were
incorporated into the final design, leading to $32 868 302 savings on the
original estimate of $71 500 948. This remarkable saving of 45% of
estimated cost was achieved by demolishing two perfectly satisfactory
buildings adjacent to the site and rebuilding them elsewhere. Within the
original scheme the design team had used considerable ingenuity to
design an expansion in the boiler house facility around the constraint of
the existing buildings.

Advantages and disadvantages of the value engineering study

A value engineering study is seen as being effective by reason of:

The generation of alternative technical solutions to a problem which
have been costed in initial and life-cycle cost terms.

The fixing of a date for the completion of a sketch design. Although not a function of the study it has been suggested that the setting of a date for the study, forces the design team particularly the engineering designers, to a more advanced stage than would otherwise be achieved.

In the majority of cases the costs of the study are a small proportion of the savings achieved. Value engineers state that on any project at least 10% of the estimated contract value is within the area of unnecessary cost. They also state that the value engineer will achieve a 10:1 return on the investment made by the client and therefore in the majority of cases the study fee, usually quoted as a lump sum, would work out to be not less than 10% of the savings realized.

In responding to questions on 'how often do studies fail?' value managers stated nil, but one major client stated about 2%.

The problems associated with the study relate to conflict, time and resourcing. These problems are discussed further in Chapter 7 but centre around:

The fact that the client may consider that the design team should arrive at the optimal solution without the need for a further exercise at additional expense. This criticism may be countered in two ways. Firstly that it is the function of the design team to arrive at a workable solution given the information in the client's requirements. Secondly, that a value engineering study is an analysis of the ideas which have been generated. A value engineering study cannot be carried out until there is an idea to analyze and it is therefore truly a second phase of the design exercise. Currently, designers are not expected to carry out nor paid for such an exercise.

The interpretation of the exercise by the design team as a critique of their design judgement. This is a difficult problem which is hard to counter unless the original designer plays a part in the activity. The reason given by some value engineers for not including members of the original design team is the danger that old ideas are defended and their presence may stifle frank comments on the design. This potential area of conflict can be alleviated through the education of the designers in value engineering techniques, informing the members of the design team at the time of their appointment that the exercise is to take place and payment of an additional fee for implementing design changes.

The time of the value engineering study. It is beyond dispute that the value engineering study will effectively take three to four weeks from the design programme. That is one week prior to the study for the distribution of drawings and information, the study week and the period of time following the study for the submission of the report, discussion and design changes. In some projects this period of time, during which the design will be at a standstill, will be unacceptable. However, in the majority of cases it is capable of being accommodated particularly in view of the fact that the study itself is a watershed between sketch design and working drawings and provides an immovable date for the completion of the sketch design.

The resourcing of a study can pose problems associated with the withdrawal of professionals from their home office for a one week period. It is a condition precedent to a successful study that members of the value engineering team are isolated from their home environment for the study period.

The study therefore is not without its problems but has consistently proved to be a very effective means for the application of value engineering.

The value engineering audit

The value engineering audit is a service offered by value engineers to large corporate companies or government departments to review expenditure proposals put forward by subsidiary companies or regional authorities. The procedures employed follow exactly those of the job plan. Following a proposal the value engineer will visit the subsidiary company or regional authority and undertake a study of the proposal from the point of view of providing the primary functions. The study may be carried out using personnel from the subsidiary company or regional authority or from another company within the group or another regional authority. The study is a global review and normally takes one or two days. It is therefore fast and relatively inexpensive.

Following the review the value engineer will submit a report detailing the primary objective and the most cost-effective method for its realization.

The audit may be criticized on the grounds that it is potentially conflict-orientated and that depth is sacrificed for breadth. However, the projects director of one subsidiary company stated that a value engineering audit on one proposal revealed a number of shortcomings with the statement of requirements to the extent that the company now adopts a policy of holding a charette before a proposal is submitted to the parent company.

The Contractor's Change Proposal

The Contractor's Change Proposal (or Value Engineering Change Proposal) is a post tender change inspired by the contractor. The United States Government include a clause in their conditions of contract which states that contractors are encouraged to submit ideas for reducing costs. If the change is accepted by the design team then the contractor shares in the saving at the rate of 55% of the saving for fixed price contracts and 25% for cost re-imbursement contracts. For example, if a contractor, on a contract of £250 000 makes a suggestion which the contractor estimates will save £10 000, then following verification and acceptance by the design team, the contractor will receive £5500. The payment is made by reducing the contract sum by £4500.

The benefit of the clause is that it allows the contractor to be pro-active and use construction/engineering knowledge to improve a facility at on-site stage.

The disadvantage of the clause is that the contract may be delayed while the design team investigate the viability of the change. For this reason changes tend to be relatively superficial.

Variations on the formal approaches to value engineering

Although the four approaches detailed above are most often described they are not suitable in every case. The following are applications of the job plan which have been used in practice but do not fall within the standard approaches.

(a) The orientation meeting. Similar to the charette, this meeting is a part of the value engineering procedure operated by New York City, Office of Management and Budget. The meeting of client representatives, design team and independent estimator is held when either the brief or the brief and schematic have been developed. The objectives of the meeting include: an opportunity for all taking part to understand project issues and constraints, to give and receive information, to determine whether all information for a 40 hour study is likely to be available by the date set for the study. A full report of the orientation meeting is given in 'The Practice of Value Engineering: Enhancing Value or Cutting Cost?', Kelly and Male, RICS, 1991.

(b) The shortened study. In many cases the estimated project value is lower than the £2–3 million considered to be the lower limit for a full 40 hour study involving a team of six. In this case the 10% rule of thumb is used to determine how much can be spent on the study. For example, for a project of £500 000 the target savings are £50 000 and the fee for the study £5000. It is now a question of determining how much professional time can be bought for £5000. If a rate of £600 per person per day is assumed, including expenses, then eight person-days can be afforded. This could be three people for two days with two extra days for the value engineer.

(c) The concurrent study. This approach involves the design team themselves coming together on a regular basis under the chairmanship of the value engineer to review design decisions taken. The method has much to commend it in that it answers much of the criticism levelled at the standard 40 hour study. The extent of involvement of the value engineer needs to be determined in advance so that the fee can be established. The fee bid from members of the design team will also have to account for their extra involvement.

The concurrent study is also suitable for construction management contracts in which the design is carried out in stages along with the construction, (fast-tracking). In a study of a $100 million office project in Canada, on which a value engineering design meeting was held on site each Wednesday, the reference point for the comparison of costs was the elemental cost plan. At initial meetings the function of the spaces were analyzed and five adjacency diagrams generated.

These were reduced to three for presentation to the client. For the selected plan a number of structural solutions were generated and analyzed on a matrix along with solutions for the electrical and mechanical installations. Once the building form was established construction work began. Meetings continued with the construction manager in attendance through to the end of the project. The final cost of the project was $9 million less than the budget.

A disadvantage of the concurrent study is the difficulty of proving the value of the time expended, it is easy to say that the budget was too high.

3.4   SUMMARY

This chapter outlines a diversity of value engineering practice in North America which is not reflected in the standard texts. It shows that practice has adapted the job plan to suit the different stages of the project life-cycle at which VE services are procured. The chapter also demonstrates some of the advantages and disadvantages of each of the various approaches. Cognizance will be taken in the final chapter of these issues when a UK method is proposed.

The next chapter outlines a series of case studies to demonstrate value engineering in practice.

CHAPTER 4 _____

NORTH AMERICAN PROJECT CASE STUDIES

_____

4.1   INTRODUCTION

The four project case studies described here are examples of the application of value engineering in a North American context. They have been chosen to illustrate the varied demands of the client and illustrate therefore a number of approaches with differing team members.

Case study 1 – Heat treatment facility for aircraft engines

This project was for a specialized building to house vacuum furnaces for heat-treating aircraft engine parts. The project comprised 12 800 ft² to accommodate the furnaces, a 500KVA substation, a new access ramp, a 2.5 ton bridge crane and an equipment mezzanine to house 2 chillers for furnace cooling.

The estimated cost of the project was $4 155 530. The client had allowed a budget of $3.5 million. The value engineering exercise was undertaken to determine whether the cost of the proposed design could be reduced without compromising function. A value engineer was commissioned on a fixed fee basis and he selected a team comprising:

- 1 certified value specialist;
- 1 mechanical engineer;
- 1 client's representative;
- 3 electrical engineers.

The workshop took on the standard format following the job plan. During the information phase a list of technical requirements were drawn up for the furnaces and a flowchart of the whole process was completed.

A FAST diagram was then compiled to increase the team's understanding of the whole process.

During the creativity phase 70 ideas were generated of which 26 were selected for development. Of these ten were developed into recommendations. The total savings on the project amounted to $1.5 million. The type of suggestions included: relocate chillers and eliminate mezzanine saving $199 395; relocate cooling towers saving $114 004; separate electrical feeds to furnaces instead of bus duct distribution saving $165 000; substitute cross-floor ventilation for in-at-roof out-at-ground saving $175 176.

This case study was prompted by an overspend but demonstrated that even where the subject for design is complex, by introducing a team of the right professional mix, savings can be accomplished.

Case study 2 – Mass transit project

This project comprised 9.3 miles of banked double track rapid transit line from a city centre to an out-of-town airport. Nine stations were to be included. The estimated project cost was $255 million.

The value engineering study, undertaken at the concept stage, followed the traditional job plan but was spread in time over four months. The total study time was eleven days. The project commenced with a one day pre-study session to advise the client and the project design team of the objectives of the study. Three months later a further one day session was committed to defining the features and the characteristics of the project. One week later a nine day study was commenced with team members made up of the proposed design team.

The team comprised:

- 2 certified value specialists;
- 1 professional engineer (the supervising consultant);
- 1 signalling and communications engineer;
- 1 transit engineer;
- 1 civil/railway engineer;
- 1 civil/structural engineer;
- 1 mechanical engineer.

During the information phase the team completed a FAST diagram showing the relationships between needs and wants from the owner's and user's perspective. Project costs were allocated to function. Owner/user attitudes were gathered during the meeting and these were related to the project functions with an indication of their perceived importance.

During the creative phase 105 separate ideas were generated and these were narrowed in later phases to 11 detailed proposals. The total saving through the exercise was estimated at $53.75 million.

This case study demonstrates the worth of carrying out a detailed study at the concept stage using the proposed design team as members of the VE study. The VE fee was $50 000, although the study was lengthy the majority of the personnel involved were the project designers or client employees.

Case study 3 – A hospital building

This case study was undertaken during the attendance by the authors and a quantity surveying practitioner to a training workshop held in the US. The project was a three-storey $24 million small general hospital, located in Northern Italy, that provided outpatient treatment or straightforward minor operations for US army personnel and their families. Requirements for major surgery were dealt with by an Italian hospital located in the vicinity. The hospital was serviced and air conditioned throughout. Exchange rate fluctuations were causing the project to remain over-budget and it had therefore stalled.

The VE team on this project comprised:

- The author (Steven Male);
- 1 quantity surveyor;
- 1 architect with health-care design experience;
- 1 contractor's representative;
- 1 electrical engineer;
- 1 mechanical engineer;
- 1 structural engineer;
- 1 building services engineer.

Due to problems within the group the team split into two analytical components; the author, the architect and the services engineer undertook a functional analysis of the project. The other team members, including the QS practitioner, undertook cost reduction/substitution/omission. The sub-group undertaking the functional analysis spent approximately one hour analyzing the project and identified the primary task of this project as to **provide medical services** where the primary function was to **organize diagnosis** since the hospital faced three sets of strategic decisions effectively determined by the initial contact between doctor and patient:

- to treat as an outpatient;
- to treat as an in-patient for minor surgery;
- to treat as a patient requiring major surgery and hence dealt with by the Italian hospital.

The analysis of the project starting from this primary task lead eventually to the identification of three primary services delivered by the hospital:

1. in-patient treatment delivered by clinical staff;
2. support services delivered by nurses and technicians;
3. administration.

This resulted in project costs being allocated on a m² basis to each of the main service types rather than to a technical specification as in normal project costing. In addition, the use of functional analysis also resulted in the project being re-configured from a three-storey heavily serviced building to a single-storey building with only the in-patient wing being air-conditioned. The remainder of the building was to be re-designed to maximize the use of natural ventilation and lighting with secondary heating support. A re-definition and re-costing of the concept provided potential savings of $8 million.

   This was compared to an initial potential saving of $2.5 million through the reduction/ omission/ substitution route.

   This case study was particularly interesting from an academic viewpoint as it demonstrated quite clearly a number of factors. Firstly, good chairmanship and control of the group is vital. Secondly, if a breakdown occurs then perhaps it is better to operate as two groups. Thirdly, the normal practice in North America of omission and substitution saved 10% of project value. Fourthly, that it is much more profitable to study concepts, even at the late stage of sketch design, if the client can accommodate the consequent time delay.

Case study 4 – An administration block

The last case study involved the analysis of a straightforward $2 million administration building for a US coastguard training facility at 35% design stage. The project in question was located in a large secure site and was planned as a two-storey steel framed building with external works to reroute an existing road. Under the original scheme the road was to be broken up and landscaped.

   The VE team on this project comprised:

● the author (John Kelly)
● 1 consultant project manager
● 1 architect
● 1 contractor's representative
● 1 electrical engineer
● 1 structural engineer
● 1 building services engineer.

   The team worked through the traditional job plan but did not compile a FAST diagram or analyze the building in terms of function. The

information phase involved working through the drawings and specifications to obtain a full picture of the project. This was done partly as
individuals and partly as an informal group without the active control of
the chairman.

The creativity session was not aimed at any particular function but
tended to wander at random over all aspects of the building. 150 suggestions were made in less than 45 minutes.

Of the 150 ideas 20 were worked up into fully costed proposals with a
life-cycle cost component. The proposals included using part of the
redundant road surface as a car park, and deleting part of the proposed
new car park. In addition, the two-storey steel frame was re-configured as
a single-storey framed structure. The air conditioning plant, originally
located on the roof, was moved to a compound adjacent to the building.
Also the internal layout was re-configured to reduce the circulation areas
and rationalize the rooms on the basis of perceived organizational
structure. The potential savings identified were $492 000 or approximately
25% of the estimated project value.

This last case study illustrates what was perceived to be mainstream
North American value engineering. Although it was effective there was a
feeling that had the function of the spaces within the building been
addressed in a structured fashion the building may well have been smaller
and the cost reduction greater.

4.2   SUMMARY

The case studies have demonstrated the following points:

(a) In case study 1, using a VE team heavily biased towards electrical
engineering in line with the project characteristics, a potentially dead
project was re-activated. Budget over-runs were subsequently turned into
savings.

(b) In case study 2, using the existing design team and a VETC, on a
complex, high value major project, potential savings of 20% were targeted
at the concept stage. This study is an illustration of the willingness of the
existing design team to be involved in a VE study and where major
savings were isolated.

(c) Case study 3 is a unique example that highlights a number of
important factors in a VE study of a stalled project. Firstly, the importance
of group dynamics. Secondly, the importance of using functional analysis

to re-orientate perceptions of the project's primary task. Thirdly, to provide a comparison of the savings that could be made by using functional analysis, even in a short space of time, and cost reduction and substitution. Finally, even although the project task was redefined, client input is essential in order to validate the correctness of the task definition.

(d) Case study 4 is a clear example of a traditional VE approach to project auditing. Although the majority of suggestions were of a cost reduction/ cost substitution nature a multi-disciplinary team review is effective.

The final conclusion that can be reached from these case studies is that the team interactions that take place during brainstorming, even in a cost reduction/substitution mode, produce many more ideas for potential savings than would be the case in a UK context. Furthermore the largest savings are likely to be made within the overall parameters of changes to the project concept.

The next chapter presents a series of case studies on the implementation of value engineering by client organizations.

CHAPTER 5 ────────────────

NORTH AMERICAN CLIENT IMPLEMENTATION CASE STUDIES

5.1 INTRODUCTION

The following three case studies of North American clients demonstrate differing implementation strategies, two of which could be said to be successful and one less so. The case studies were undertaken using interview techniques on two visits to North America during September 1986 and April 1989.

5.2 CASE STUDY 1 – US NAVY VE PROGRAMME

Outline of the Navy VE programme

The original impetus for VE studies came from Naval Engineering who applied value engineering to the ship building programme. The initial implementation of the VE process was from top management down and the maintenance of this approach has resulted in its continued success.

The US Navy has a large portfolio of existing property of which approximately 90% is over 30 years old. There is a significant annual building programme usually 2000 or so projects with a combined value in the range of $1.2 to $1.5 billion. During 1988 the Naval Facilities Engineering Command commissioned over 2000 architect/engineers to design over 2300 projects with a combined value exceeding $1.6 billion. A large proportion of the projects relate to repair and maintenance and these are not subject to value engineering nor are they accounted for in any meaningful way. Any project over $2 million is subject to a VE study the limit being chosen because approximately 80% of project construction values occur above this point. The money gained by savings through VE is transferred into other building projects.

The administrative costs for the VE studies exceed $4 million due to the increasing use of studies carried out by contract. There are only two VE staff, the VE management group, in the Navy who can call on manpower within the organization if they are needed. In-house people are used to review projects or to participate in studies and their costs are additional to the total figure quoted for the cost of administration. Included in the administrative figure is the cost of re-design work. This can vary in any one year from a few hundred thousand to over a million dollars. The Navy have a policy of paying for re-design work consequent to a VE study.

The VE management group within the Navy accept that the Navy spends more on projects than outside organizations because of: the Davis/Bacon Act, the Equal Employment Opportunity Act, the Small Business Set Aside Act, the Contractor Quality Control Act and other similar legislation. The Navy in common with other public services, is subject to open tendering in which the lowest bid must be accepted. Therefore, if a contractor can get bonding he can bid on and win a Navy contract.

The navy is also subject to 'The Federal Acquisition Regulations' which requires that every contract over $100 000, regardless of how the funding is obtained, has to have an incentive clause whereby the contractor gets 55% of the savings. An incentive clause will be used in a project which has also been the subject of a VE study, however, contractors tend not to contribute anything meaningful. Electrical contractors appear to have the most success with the incentive clause although there was no reason for this. The incentive programme generates less than 3% of total VE savings per year.

VE consultants

VE services for the Navy are provided by a second team of designers, separate and completely independent from the original designers which prepare the 35% plans and specification. The second team of consultants performing VE Team Studies (VETS) must meet the following requirements:

1. The team leader must be a current Certified Value Specialist (CVS).
2. All members of the team shall be professionally registered architects and engineers with at least 12 years experience. They are to be completely knowledgeable of VE methodology by attending a certified forty-hour study.
3. A list of a minimum of six team members and their respective CVs representing the various disciplines to be covered together with the certified (CVS) team leader's qualifications and discipline shall be submitted for approval at the time of negotiations. Changes or

substitutions to the approved VE team configuration shall be submitted in writing to the contracting officer for approval.

4. Team members must not have been associated with the original project design team. When preparing the fee for VE services the team leader, the VE contractor, is required to hire three of the six team members from local architect or engineers practices but no members of the practices of the design team representatives may be a member of the VETS Team.

5. The VETS team will perform the study either in the town where the architect or project design engineer's office is located or adjacent to the project site. The VETS team are isolated from their normal work station in order to avoid the normal daily interruption such as: phone calls, quick questions and brief meetings, which come up and tend to be very disruptive to studies of this type.

6. The team leader must deliver the VE report no later than seven days following the study.

7. The VETS team is to meet on five consecutive working days (40 hours).

8. A typical value engineering team and the anticipated input from team members is as follows:

(a)	VE team leader	80 h;
(b)	architect	40 h;
(c)	structural engineer	40 h;
(d)	mechanical engineer	40 h;
(e)	electrical engineer	40 h;
(f)	civil engineer	40 h;
(g)	typing	60 h;

VE team study

The VE study is a 40 hour five day programme for a multi-disciplinary team usually comprising six people although team numbers may increase or decrease depending upon the complexity of the work. There are between 150 and 250 studies undertaken each year costing approximately $20 000.

(a) Design briefing is preceded by a pre-design conference as a result of which a preliminary engineering document is produced. Once a consultant is under contract the VE group will have a joint meeting with the lead architect or project manager from the Navy, the customer and the design consultant at which the information contained in the briefing document is discussed with specific reference to how the customer plans to own, generate and maintain the building. Once a concept for the project is agreed the design architect will proceed to schematic design.

The Navy have produced a few VE exercises on user requirements but generally very little work is undertaken in this area although it is accepted that the preliminary stages are where the greatest amount of money can be saved. A problem is seen to be that there is very little cost information and high-cost areas are not known.

(b) The 40 hour VE study will follow the five-step job plan as recognized by the Society of American Value Engineers (SAVE). The VE Report (15 copies) shall encompass the recommendations of the VE study group with detailed cost estimates, life-cycle analysis and sketches, as necessary. The Navy require that VE services are performed in a timely manner concurrently with the normal design procedure and without delay in the design schedule set forth in the architects and engineers scope.

It is not normal to undertake VE studies beyond schematic (35% design), however, VE studies are also undertaken at production drawing or tender stage (90% or 100% design) if the project is in financial trouble. FAST diagrams are not normally constructed in 35% studies but in concept studies they can be used to determine exactly what is to be accomplished. At 35% design, the requirements have been identified fairly well.

The navy VE group consider that enthusiasm is important in generating VE savings and without this the VE study will fold. VE consultants are trained, they want to do studies, they are in the VE business and they have the enthusiasm for it. Internally, staff may not like to do VE and therefore very poor results may be obtained. The group consider that a VE consultant will develop about 40–50 recommendations whereas an in-house study will generate about ten. The knowledge of in-house staff is good but they do not have the same enthusiasm because of the heavy workload waiting for them back at base. To consider using any other approach is believed to be a waste of money.

(c) A resolution meeting is arranged for approximately 3 weeks following a study and a spreadsheet is made up which gives a description of the project: recommendations and an initial estimate of savings. Comments are sought from:

- the Navy VE group;
- design consultants;
- user, ie., the customer.

VE recommendations are judged by the use of the matrix sheet with recommendations on one axis and the members of the resolution meeting on the other axis. Prior to the meeting each member enters 'yes' or 'no' against each recommendation and about 60% of recommendations are decided in this way. If all parties accept the recommendation then there is no need for discussion. The meeting is therefore held principally to discuss those recommendations that have not been accepted by everyone. Reasons are sought as to why the recommendation is rejected and often this was because all the relevant information was not available. Open discussion at the resolution meeting is very important.

In terms of design liability and VE recommendations, if a VE recommendation is rejected by the consultant for a weak reason then the Navy will override him and seldom is a liability problem encountered. If one is encountered then the Navy, as client, are prepared to take liability themselves.

The VE group have found that generally the consultant who has been VE'd before does a lot more fine tuning of the design and after experience of three or four studies it becomes more difficult to find savings. If the drawings are returned to the design consultants with a lot of recommendations then this could be an implied criticism, however, this is not intended and is out of context with respect to the processes and objectives involved.

The Navy has had VE failures but these are a small proportion of the VE studies undertaken, less than 2%. They have had plenty of recommendations that have failed because either the project manager was weak or there was a very tough customer. They have had return on investment of implemented suggestions as low as 1.5% but normally they would expect 25% or better.

Cost data and exercises

The initial cost estimate is built-up from the Navy cost data-base and the designer's initial scheme. The consultants send the quantities to the Navy who produce the estimate, adjusted for geographical factor and for difficulty. An LCC analysis will be undertaken during a study based upon data derived at that time. The Navy do not keep a life-cycle cost data-base which the VE group acknowledge as a large shortcoming since an LCC analysis requires historical data. Energy cost data is also not kept although recently individual buildings have been metered for energy consumption. The VETS are permitted to examine staffing costs, in fact the VE team has *carte blanche* to save the Government's money.

5.3   CASE STUDY 2 – NEW YORK CITY VE PROGRAMME

Outline of the New York City VE programme

Construction projects in New York City are proposed by an agency which will be responsible for their operation and maintenance e.g.

Department of Environmental Protection (DEP). Each agency receives a part of the annual budget and requests are received by the Office of Management and Budget (OMB), New York City (NYC), for funding on a list of projects.

This case study outlines the introduction and implementation of value engineering into the procedures of OMB. The purpose of OMB is to review the capital programme of NYC. The unit of OMB which incorporates VE comprises 18 staff of which 10 are project managers whose responsibilities include monitoring change orders and reviewing agency standards as well as participating in the VE programme.

In a study preceding the setting up of the VE procedure, it was found that 81% of capital expenditure was committed to only 12% of the 3900 projects which represented the city's construction programme. The high cost projects were targeted and a programme was set up to value manage about 500 projects over a period of four years.

An OMB budget task force is appointed to monitor an NYC agency. This task force is the strongest ally of the VE group within OMB. The OMB budget task force will review the capital programme and agree with the agency concerned which projects are worthwhile and can proceed for funding and which should be deferred. The agency's priorities within the agreed capital programme then become the governing factor.

One of the functions of an OMB task force is setting the budget for an individual project. However, project budgets are frequently found to be unrealistic because they are made prior to any preliminary studies being undertaken or the appointment of designers.

OMB adopt a life-cycle approach to the evaluation of their facilities with a time horizon of 20 years, the bonding period for finance. In the life-cycle cost calculation no assumptions are made for improving technology in the future and the calculation is based on the most efficient technological means currently available. Maintenance and refurbishment costs are assumed.

The cost of a VE exercise, including a contingency for additional meetings or other unanticipated expenses, is covered by an OMB task order which is approved by the budget director.

New York City's first VE exercise was carried out in 1982 on a gaol construction project on which serious budget cost overruns had occurred during early design. VE was suggested as a cost control technique in response to a directive from the budget director to reverse the escalation. OMB invited presentations from consultant value managers and subsequently decided to commission a VE exercise which was an outstanding success.

This first study was of three days utilizing a small group of professionals in a VE exercise. The size of the building was reduced by 25% which was agreed to by the Department of Correction, the Department of General

Services and the architect. As a result of the re-design, 66 full time staff positions were dispensed with which was worth more than the capital cost savings. The project started $11.1 million over budget and after the value engineering exercise was $4.4 million under budget.

Subsequently VE was tried on another prison project and found to be even more successful. Slowly a value engineering programme began to evolve, starting with small projects such as the computer networking system. This project was prompted by the amount of money the city was spending on telephone charges to support a communications network throughout its various agencies. This particular project was seen as a challenge as the VE office knew very little about computers.

Value engineering was later applied to a broad variety of very different projects and these were all found to be successful. However, notwithstanding this, the prime reason that value engineering was initiated results from OMB's position as the funding authority which is concerned with the efficient disbursement of the city's money. This is seen as both minimizing expenditure without diminishing quality and guarding against spending a large sum of money only to find that the project does not achieve the required function.

Design team consultants

The designers of the facility are employed by the commissioning NYC agency. The designers sign the drawings and are responsible for all design decisions. OMB make suggestions which they feel are feasible and viable and have them agreed for incorporation in the design. Normally the designers commissioned by an NYC agency do not know that they are to be the subject of a VE study at the time of their appointment. However, the participation of these designers is encouraged in the VE process so VE is treated as an additional recompensed service. The agencies are provided by OMB with a sum to cover the hourly costs of the designers participating in VE namely, presentation, attendance at two or three meetings, and reviewing the study responses.

Some designers have experienced VE previously but not in the structured form conducted by NYC. Some have experienced VE conducted by construction managers, which, in this instance, are constructability reviews and tend to be more 'amputation' than functional analysis.

OMB never compromise function and once designers have understood this resistance becomes less. OMB appreciate the threatening nature of the VE process and attempt to reduce it. However, designers appear to lack this appreciation but tolerate and learn from the process.

Delays in decision making by NYC is recognized by OMB as the major problem and source of criticism from designers who wish to expedite the project. OMB have attempted to reduce delays at agency level but acknowledge that delay is the major area where they are exposed to criticism.

VE consultants

A value engineering team co-ordinator (VETC) is appointed on a two year contract and will become a member of a pool of seven VE firms which are appointed on a rotating task order basis. At the time of the pre-quali-fication bid, VE firms are asked to submit proposals for design consultants to be VE team members for:

- A standard team for a typical building study;
- A team for a complex building study;
- A team for an infrastructure study;
- A team for a training study.

The pre-qualification proposals are for specific people from named firms who are viewed as sub-contractors to the VETC. When VETCs are instructed to field the standard team, OMB would expect those described in the tender submission to undertake the study but will allow substi-tutions and specialists where necessary. Where more than one study is undertaken on the same project OMB would expect to see the same consultants at each.

The fee bid is one part of a tender procedure which also includes an assessment, through OMB's experience or by references, of such factors as:

- Quality of work;
- Responsiveness;
- Flexibility;
- Timeliness in the studies and in the delivery of reports;
- Ability to work well with designers;
- Competitive or collaborative;
- Previous client's observations;
- Strengths and weaknesses;
- Ability to motivate;
- The proposed team composition.

In respect of qualifications for value engineering NYC specify CVS or equivalent on the basis that some good people are not CVS. However the argument is also put forward that if these people are going to make VE their career then they should become a CVS.

At the time of the project the VETC will return a further fee bid with proposals for team members indicating any substitutes and specialists.

Team members and specialists are paid by the VETC from the VE umbrella contract. OMB insist on an independent team rather than the existing design team in the VE study on the basis that it is difficult to be critical of oneself.

In addition to the VETC and the team, selected as above, NYC appoint an independent estimator. The estimator is chosen from a rotational list of 15 estimators pre-qualified for city estimating work. These estimators comprise cost consultants and quantity surveyors including a few British firms. The list of 15 firms is given to the VE consultant who asks three to four to take the project description and put in a fee bid appropriate to the stage in design. For example at design-brief stage it would be a square foot estimate, at design stage a more detailed estimate. The fee bids are sent direct to OMB who choose the lowest acceptable proposal from the three. The estimator is therefore employed direct by OMB and has no contractual link with the VETC.

VE study

Each agency identifies candidate projects which are tracked and assessed for a VE study. Examples of typical projects are schools, school pools, school zoos, an aquarium, subway stations, water pollution control facilities, gaol, fire houses and police stations, incinerators and large highway reconstruction. Generally, projects are selected which meet all of the following criteria:

1. Projects of $10 million or more.
2. In the capital plan for one year ahead.
3. In brief development stage or at the point of designers' contracts.

The project is targeted and then monitored. The sponsoring agency is informed that the project is to be the subject of a value engineering study. The monetary bottom limit on a project is determined by the requirement to make a minimum 10:1 return on investment. A NYC VE study costs $50–$60 000. For a minimum amount of expenditure on the study, OMB can achieve savings on budget normally of between 8% and 15%, within a range of 3% to 30%. At times omissions are made in the VE study which subsequently have to be added back and in this unlikely event the overall savings are so great that the additions can be easily funded.

(a) At the pre-brief stage there is a long period of project 'tracking' and intelligence gathering by OMB prior to the initiation of a value engineering exercise.

A NYC agency will make a 'ball-park' estimate based on a similar facility elsewhere in the USA. This cost is factored up for location and time. The agency will make a Request For Proposals (RFP), the first stage of the briefing process when design consultants are commissioned and the design contracts are approved by OMB. The OMB task force commences monitoring the process.

As projects come on-stream at RFP stage the next firm of VE consultants on the rotational list is sent a notice of assignment and informed:

1. That the project is on-stream;
2. The nature of the project;
3. The standard team members (by discipline) needed to review the project;
4. Any specialist which OMB feel is required.

(b) At the brief or fully schematic plus brief stage a value engineering orientation meeting is undertaken. In large or complex projects, usually exceeding $50 million, the orientation meeting is self-contained and is solely concerned with the brief. In other projects the orientation meeting precedes the full 40 hour workshop and works from a data-base of schematics plus brief. In the view of OMB representatives most value engineering exercises concentrate on technical solutions, for example, the use of structural steel or concrete etc., but

'the bulk of the opportunities for savings are before you begin design. You make sure that you build what is needed not what people think that they need'.

The aim therefore is to examine the project requirements in terms of function and the efficiency of achieving these requirements.

At the orientation meeting, chaired by the VETC, the designer gives a brief description of intent but not a full design presentation. The orientation meeting will achieve a number of objectives:

1. Make sure that everyone involved in the project understands all the issues and constraints.
2. Provide everyone who is to make a decision an opportunity to give and receive information.
3. Provide the estimator an opportunity to understand the project. The estimator will commence the estimate immediately after the meeting, receiving all necessary documents at the meeting or immediately after.
4. To determine whether all the information is available for a 40 hour VE study. For example, a study could be called-off where construction costs are available but operating costs are not. The VETC has an opportunity to discuss the VE methodology and to acquaint people

who are unfamiliar with it of their responsibilities and roles in the process.

5. To examine the information to determine the intention of the project and the scope of the VE study. Emphasis is given to the running costs and therefore the VE study team members must reflect the knowledge required. 'Its the real-time opportunity that you have to affect the operating costs forever'. The information required may include staff organization, union agreements etc. The VE team must include members with experience of operating the type of facility under examination.

6. It allows OMB and a NYC agency to steer the designer towards a solution which is compatible with NYC's objectives, for example, lowest possible operating costs.

(c) A 40 hour value engineering study immediately follows the orientation meeting on a typical project. However, on larger or complex projects, usually above $50 million a two-phase process operates. An orientation study on the brief, followed by a 40 hour value engineering study dealing with concepts, and occurring between 14 days and two months after orientation. At the study the designer makes his presentation in depth and information is provided by the agency and other sources on the first day, usually a Monday. The remainder of the week follows the standard study pattern.

A report is issued following the study and a week or two thereafter, the design team will respond to the agency. The agency is required to send written responses to the recommendations in the report on the basis of:

● Accept;
● Adapt, improve or modify;
● Reject for a good reason. In this case the agency has to state the good reason;
● Further study where the idea has to be considered later in design;
● A disagreement response where a VE idea makes technical sense but goes against operating standards. In this case OMB may challenge the standard and take this to a higher authority where the matter is resolved quickly.

(d) An implementation meeting is held within a month following the 40 hour VE study at which all the decision-makers come back together and discuss the ideas and design team and agency responses. All issues are resolved at this meeting and this would signify the end of the VE process for the majority of projects. For projects in excess of $50 million the design

may be reviewed at an additional study and this has happened on gaol and hospitals, where in the case of the latter there have been up to four studies.

5.4   CASE STUDY 3 – DEPARTMENT OF PUBLIC WORKS, CANADA, VE PROGRAMME

Introduction

Identification of the potential benefits of VE to building programmes occurred in 1978 when a Senate Committee was reviewing the operations of the Department of Public Works, Canada (DPWC). The committee were aware that VE had been applied by the US Navy and recognized a potential within DPWC. This led to the formation of a group, under the direction of a project manager, to investigate the potential for the implementation of VE and as a consequence, research studies were commissioned to review the operation of VE in the US.

The study group, operating from headquarters, reported back to the Executive of DWPC recommending that the VE concept be introduced broadly into the operating levels of DPWC.

Structure of the Department of Public Works, Canada

DWPC has a decentralized regional structure comprising a headquarters staff and six regional offices in which approximately 170 project managers are employed as main project officers. Their primary function is co-ordination between departments and to act as an interface with design consultants.

The Department uses a method, or set of standard operational procedures, for project procurement called the Project Delivery System (PDS) which, in theory, is to be used on every job. Originally, the PDS had made no mention of either cost control or VE but in the early 1980s a study was undertaken to update the method to include VE as an addendum. The study concluded that a VE exercise should be undertaken at the schematic-design stage.

Appointment of design consultants

Design consultants are chosen from an inventory maintained by DPWC. Consultants are engaged on DWPC consultant Forms of Engagement which have been developed in close co-operation with professional bodies and are continually revised.

Cost consultants in Canada are retained by the primary consultant, an architect or engineer. The cost consultant has no direct contact with the client as Department policy is to have a single point contact and contract with the prime consultant. However, due to the decentralized nature of DPWC this procedure may be varied by the regional offices.

A structured appointment system, based upon fee level, operates which restricts regional autonomy. If a consultant's fee is below C$15 000, regional offices have the authority to appoint direct. If the fee lies in the range of C$15 000–C$100 000 the Minister appoints direct. Fees that occur over C$100 000 are subject to a DPWC procedure called ROUTE. The Minister approves a minimum list of 3 consultants who are asked to prepare a proposal on how they would undertake a project. This is assessed and rated by a panel prior to approval.

Appointment of a VE consultant

One of two methods could be used for the appointment of a VE consultant. Either the normal Form of Engagement for a consultant could be used, or alternatively, if the VE consultant is considered to be of cost consultant status he could either be hired direct by an architect or under direct contract as a 'sub-professional', ie. not having full consultant status. In this latter instance, a personal service contract would be used.

Developments towards the implementation of value engineering

Following the activity of the study group described above a consulting firm was hired to stage a one week VE Workshop to be conducted at headquarters. The package was delivered at four separate sessions, one in 1980 and three in 1981. These sessions were delivered mostly to DPWC staff but external architects and engineers were also invited. These workshop seminars were, in the main, targeted at people from the regional offices and acted as an internal gestation point for VE.

Although the Deputy Minister, who was head of the Department's Executive Branch at the time encouraged the adoption of VE a directive was never issued by headquarters and regional offices could only be said to be 'nudged' towards its implementation.

Suggestions were made for conducting pilot studies and these were undertaken at regional level due to the marketing effort of a VE consultant. Subsequently, headquarters attempted to generate a co-ordinated effort with the VE consultant to offer his services to regional offices.

Headquarters monitored and encouraged the use of VE as much as possible but it was not used extensively. In those cases where it was used, each VE study was unique and there was little opportunity to develop standard procedures. Therefore, in practice, VE was nurtured at the centre with policy recommendations but unless regional offices wanted to use it then implementation did not occur.

Difficulties encountered in implementation

The case study highlighted a number of difficulties in the implementation process and raised a number of important issues.

1. VE was seen as being within the scope of the architect's function as opposed to anyone dealing with cost-related issues.
2. VE was introducing change and changing traditional forms of practice.
3. VE was seen as another form of what should be happening anyway.
4. Concern on the amendments necessary to the standard terms of engagement for architects and engineers and any resulting contractual issues that could arise from re-design work as a direct result of VE recommendations.
5. In-house project managers, as opposed to designers, were the focal point of the introduction of VE. It was they who attended the VE workshop seminars. VE was therefore, from the outset, attributed to the PM group, one essentially responsible for a liaison and co-ordinating function without direct design responsibilities.
6. DPWC were, at the time of the research study, in the process of re-organization and the implementation problems that were being experienced with VE could, however, not be disassociated from changes that were occurring in the Department generally. VE was seen as just one more element of a more diffuse series of change events. Furthermore, at the time of conducting the research study VE did not have a high priority in terms of the changes that were being introduced and because of this it was dropping further into the background.
7. The decentralized nature of DPWC and the necessity for an array of projects to be carried-out in order to develop potential standard operating procedures meant that the codification of VE procedures presented great difficulties. Further, the discretionary powers of the regional offices resulted in *ad hoc* approaches. It was stated that for a full implementation process to occur outside pressure would be required, probably in terms of a renewed political initiative. Additionally, any global implementation impetus would benefit from a perception that the private sector were actively adopting VE procedures.
8. At the time of the research DWPC held no central cost-data for use in initial or life-cycle cost studies. These were seen to be necessary for a VE programme.

It was concluded that if VE was to be implemented in DPWC then a method would need to be established for the employment of external VE consultants. Also a strong 'top-down' directive was seen to be essential to overcome prejudices and fears among individuals and some operational groups. At the time of the research it was seen as being 'a good idea' by senior management and was also enthusiastically embraced by a few individuals however, no overall structure was in place.

5.5   SUMMARY

The three implementation case studies demonstrate different approaches to the implementation of a VE methodology. In the case of the US Navy it was a clear directive from senior management following its successful application in the ship building division. A VE programme was set up with clearly defined procedures.

In the case of New York City implementation was born from a demand by the treasurer's department to reduce costs on a single project. A member of staff in the treasurer's department was given the task and authority to investigate its potential and subsequently to set up a working procedure. Similar to the Navy these procedures are clearly documented and well understood.

In the case of the Department of Public Works, Canada, the existing management structure of the department acted against the institution of entirely new procedures. Revision of the Project Delivery System is necessary before the regional divisions of the department are required to institute VE procedures. Therefore while there is an appreciation of the benefits a directive from the department executive appears unlikely. Further, the department was going through a period of substantial organizational change and VE was seen as one more facet of this. This factor mitigated against its likely adoption.

For a successful implementation it appears that clear authority is required from the top management of any organization to allow someone with enthusiasm to institute a clearly defined programme. VE does not appear capable of implementation through the consensus of many diverse departments and professionals.

The next chapter provides an insight into a major North American value engineering consultancy.

CHAPTER 6 _____

A CASE STUDY OF A NORTH AMERICAN VALUE ENGINEERING CONSULTANCY PRACTICE: SMITH, HINCHMAN AND GRYLLS, WASHINGTON, 1986

6.1 INTRODUCTION

This chapter is a record of a detailed study reinforced with interviews of the consultancy practice, Smith, Hinchman and Grylls. The study was undertaken during September 1986. Smith, Hinchman and Grylls was chosen as the largest value engineering consultancy specializing in construction in North America which offered a wide range of value engineering services.

6.2 ORGANIZATION STRUCTURE

The consultancy group are architects, engineers and planners with a staff of approximately 400, of whom 300 are resident in the head office in Detroit. The group has grown by the acquisition of other organizations, eg., landscape planners and prison consultants. The Washington office, the Value Management Division, was established in 1975 through acquisition and forms one small part of the group. The Value Management Division is accepted as being one of the largest value management consultancies in the construction field.

6.3   CLIENTS AND WORKLOAD

The range of clients are indicated below:

- 15 government agencies;
- 30 municipalities or state governments;
- 40 private corporations;
- 20 overseas client bodies.

The 1985 workload of the Value Management Division was as follows:

- 30 value engineering studies;
- 10 life-cycle studies;
- 20 estimates and cost models;
- 10 programming studies;
- 5 post occupancy evaluations.

Completed value engineering studies include:

- 150 education and commerce;
- 75 environmental;
- 50 health care;
- 25 industrial process;
- 50 transportation and other.

The practice obtain work by reputation and by marketing. Architects generally have been very positive and a number of architects have commissioned the practice to carry out value engineering studies. The practice claim that value engineering, as practised by them, does not delay the design process.

6.4   SERVICES

The consultancy group can offer full services in-house with all capabilities and all specialities. For example, structural, mechanical and electrical engineering and also landscaping and prison expertise.

The Washington office offers value engineering, life-cycle costing, estimating, design programming and post-occupancy evaluation services and will take contracts in any of these fields. However, the Washington office does not offer a traditional design service.

The consultancy practice provides separate services for brief writing and value engineering studies during the briefing phase. The main emphasis has been on the latter. However, the Washington office is moving increasingly into offering brief writing as a service. (In the US brief writing is termed 'programming').

The practice is prepared to investigate the briefing concepts in some depth if this is required by the client, and indeed consider that this is necessary before design commences. The practice believes that the use of value engineering procedures at the brief writing stage enables the complete understanding of the project. It is quite common to find that clients refine their project requirements following this exercise.

6.5   VALUE ENGINEERING SERVICES

The practice undertake the full range of value engineering services described in Chapter 3. The particular type of study will depend upon the requirements of the project and the demands of the client. On a large project the practice will undertake a study at the briefing or schematic phase, at design development and perhaps at working drawings. This means they will get two or three opportunities for evaluation. The value engineering team will remain the same for each stage which enables a greater capitalization on project knowledge during the later studies.

Where the form and function of a building has a staff significance the practice like to study this, especially at the briefing stage, but only if it is within the scope of the work as set out by the client. The practice like to undertake a life-cycle budget as a part of the brief preparation and will prepare a budget for the first year and if required, for subsequent years. These budgets will include staffing costs. Hospitals, prisons and waste-water treatment plants are good examples of projects where life-cycle staffing costs are determined by the design of the facility. The practice do not often get involved in the staffing costs of offices.

40 hour value engineering study

The practice have 50–60 people trained in study procedures and use them repeatedly in the 40 hour studies which is the practice's standard procedure for implementing value engineering. Generally, all members of the study will have experience of value engineering, however, one person may be brought in who has no experience.

Specialists are used on a regular basis to supplement the team, resulting in a value engineering team which is unique and reflects the demands of the job. Specialists may include, for example, hospital consultants, and/or transportation consultants. They may also hire in estimating services where they do not have sufficient resources in-house or where the cost of travelling expenses will mean that it is cheaper to hire expertise on location. The practice will use the yellow pages, journals or references from someone else as a source of specialist expertise. Sometimes the owner/

client will be on the team or even the architect/engineer. The practice do not object to having the architect in the room participating, indeed this can be an advantage since the architect will have all the necessary contacts for data and computations.

The value engineering team co-ordinator (VETC) will construct a FAST diagram either where the client requires it or where the study is of a process. In view of the consultancy the building may be construed as a process.

The practice believe that the VETC cannot handle both the job of VETC and cost consultant in a VE study and therefore, a cost consultant will be employed as a part of the VE team. Sometimes the client will demand an independent estimate and in these cases or when the project is undertaken overseas a cost consultant, independent of the team, will be hired to carry out this task.

High cost areas on a project will be identified by first creating the cost model in UNIFORMAT elemental form. The VE team will then do cost *versus* worth analysis in major systems. The VE team will also analyse the project in terms of parametric quantities.

The practice obtain the cost data for VE study from:

1. the estimate, which is developed into a cost model;
2. for worth data; from experience, comparables, history and past projects;
3. means or similar price books;
4. practice experience and historical data.

On a five-day study where the architect has not been involved, the practice like to have the architect in on the Wednesday of the study to go over the brainstorming idea list. The architect will be asked to help rank the ideas and evaluate them. The architect may then indicate that a particular idea has been considered and rejected for a reason. The VETC may accept this but may still respond if the idea is altered and the reason for rejection is no longer valid. The practice do not like to spring recommendations on the architect on a Friday afternoon (the text book approach). 'This is deadly'. The practice admit there are some VETCs who do this.

(a)   Preparation for a 40 hour value engineering study

The practice will normally undertake the following prior to a value engineering study:

1. Co-ordination of the work;
2. Gathering of data comprising a project summary describing and highlighting major project considerations, eg:

(a) site condition and site investigation reports, project constraints and operational requirements,
(b) planned construction schedule and date for completion,
(c) plan layouts,
(d) local design and material standards,
(e) site and general layout drawings,
(f) mechanical and electrical drawings,
(g) design, criteria and calculations for major units, sub-systems, structures and buildings,
(h) design criteria for administration, storage, maintenance, employee facilities, roads, parking, plus
(i) vehicle storage and maintenance,
(j) estimated energy demand at average and peak periods including material and/or backup calculations,
(k) estimated construction cost for sub-systems including backup cost estimating worksheets with quantity takeoffs,
(l) estimated annual operations and maintenance costs with backup worksheets broken down into the same categories as the construction costs,
(m) estimated costs and frequency for major equipment and components requiring replacement during the planning period;
3. Scheduling of study location and team members;
4. Preparation of data (duplicate sets);
5. Modelling of energy, space, initial cost, life-cycle cost and, if possible, time. The practice prefer to prepare the above models before the study. The cost model is always prepared but sometimes it is not possible to obtain the data to prepare the others.

Value engineering audit

For one particular large corporate client the practice have, for the past three years, undertaken a design audit. They audit the whole design process ie., the capability of the people doing the work and the capability of the divisional management that ordered the work. The practice's function is to make sure that nothing is forgotten ie., that the major participants have thought of everything relating to the project. An example of the factors considered are:

• how the raw material comes in the front door and the finished product goes out the rear door;
• layout and arrangement of space;
• worker productivity and environment;
• human factors of comfort and amenities.

The corporation wishes the audit to be a part of their corporate strategy and any architect involved on one of the corporation's projects should be aware that an audit programme exists. The practice consider this to be another form of value engineering because of its relationship to value. Often the audit is done before the appropriation of corporate funds to the subsidiary company.

In the audit the practice will also comment on the project image, a situation where the level of the building cost is almost secondary and when the corporation is more interested in the qualitative aspects of the project. The completed report goes to corporate headquarters.

Fees and terms of engagement

The practice do not use a fee scale for value engineering work. They will estimate the fee for a specific job or will propose a standard schedule of service in terms of man hours and labour rates, travel and *per diem* expenses, reproduction costs, overheads and profit. The range of fee comes out at between 0.1% and 0.5%. One client asked for drawings after a study and the practice provided these for an additional fee.

The value engineering team co-ordinator is appointed in two ways:

1. By contract through the architect/engineer where the architect selects the value engineering consultants.
2. Independently. In this case the architect will be told that a value engineering study will be undertaken. The AIA standard form of appointment states that architects should be aware of the policies of the companies that they are dealing with.

Sometimes the client sends his own conditions of appointment to the value engineering team co-ordinator but more usually the practice's own model conditions of engagement are used.

The practice operate a very fast turn-a-round for a VE project. A study can be researched, undertaken and reported within a three week period i.e., 15 working days.

6.6   SUMMARY

This is a case study of a corporately successful consultancy offering VE services. A number of points emerge:

1. The consultancy has a broad and in-depth knowledge base internally on which to draw for VE studies;
2. They tailor their VE services to meet client requirements and are also prepared to be innovative in service provision;

3. They have developed a standard operating procedure for VE studies;
4. Three of the senior executives are recognized internationally as being at the forefront of value engineering development through their textbooks.

This chapter concludes Part Two, the North American implementation of value engineering. Part Three is a critique and analysis of the earlier chapters, drawing together the insights into an exploration of important issues for consideration in a UK context.

PART THREE ———————————

AN ANALYSIS, CRITIQUE AND EVALUATION OF NORTH AMERICAN VALUE ENGINEERING

———————————

CHAPTER 7 ———————————————

A CRITIQUE OF NORTH AMERICAN VALUE ENGINEERING

7.1 INTRODUCTION

This chapter provides a critique of North American value engineering based on research undertaken by the authors using case studies, interviews and questionnaires. In comparison to North America, the presence of the quantity surveying profession in the UK construction industry is one major important difference in the manner in which cost consultancy is handled within the two countries. As indicated in Chapter 3, in North America costing is undertaken by estimators employed directly by the architect. Therefore, project costing is a much more passive approach than in the UK, with an estimated out turn cost being the consequence for a project rather than cost monitoring or cost control.

Within this context, there is no doubt that much of what purports to be value engineering in North America is the equivalent of many of the functions provided in the UK by the quantity surveyor. For example, when the authors attended a value engineering training workshop many of the items targeted during a brainstorming session for VE recommendations could be subsumed under the headings of either cost reduction of component/ material substitution. However, it also cannot be denied that the presence of a multi-disciplinary team added to this form of project appraisal since items were being thrown-up for consideration by the team that may not have emerged without intra-group exchanges during a free flow of ideas.

Furthermore, inspection of North American project cost estimates by the research team also revealed considerably more hidden contingencies than would normally be expected in the UK. In some instances this ranged as high as 25–30%, whereas an equivalent figure in the UK would normally be around 5%. This suggests also that any project appraisal undertaken as a value engineering study at 35% of sketch design on North American projects would have scope to find savings anyway. Put another way, there appears to be considerably more 'fat' in North American cost estimates to be expunged than would be the case on British projects due to the ongoing involvement of the quantity surveyor as a cost consultant in the design team.

There is also a heavy public sector patronage of value engineering studies in construction, suggesting a close linkage with public accountability. Interviews conducted in Canada and the US suggested clearly that diffusion of value engineering into the private sector was slow and required considerable marketing effort on the part of practitioners. There is also considerable evidence to suggest that heavy public sector patronage, especially that of the US Navy, has established the guiding principles of VE practice, namely, the conducting the 40 hour workshop at 35% design.

7.2 PERCEPTIONS OF VALUE ENGINEERING

The investigations of North American VE by the research team also involved ascertaining the perceptions of value engineering from a sample of contractors, architects, engineers and delegates attending a training workshop. This was undertaken through the use of questionnaire surveys. The results of these surveys revealed a diversity of views.

Definitions of value engineering

Delegates attending a VE training workshop were asked to define value engineering in their own terms. The major thrust of their answers were related to cost reduction or cost optimization in some form. A sample of definitions also provides additional insights:

> 'value engineering is a method to reduce project cost by changing materials or combining material changes with architectural or engineering changes'.

Another definition suggested the following view:

> 'value engineering is a method of improving cost, schedule, constructability, without jeopardizing system integrity. It allows constraints to be challenged. It gets back to basics and allows the owner (client) to judge the basics.'

Other examples of definitions include:

> 'an integrated process of design and construction that produces the most appropriate building to the user's criteria'

'an evaluation of alternative construction materials and systems to save money without major effect on programme, maintenance or appearance, chosen on a priority basis'

'a systematic and creative means to influence positive and profitable change through small group interaction in a relatively short time span'

When questioned further on the problems of implementation of value engineering, respondents provided a wide range of feedback. However, the following points summarize the main themes:

1. Being able to convince the client or architect/engineer to allow a value engineering team to participate in the design and costing phase before the project plans and specifications are completed;
2. Being able to achieve the appropriate building at the right price;
3. Clients being convinced that savings can be made;
4. Value engineering is likely to have a low priority in the client's or design team's thinking when they are faced with deadlines.

The next section deals with the timing of value engineering studies.

7.3   THE TIMING OF VALUE ENGINEERING

The authors were able, through assimilating case study material and interviews with practicing value engineers, to ascertain a wide range of target points for a value engineering study. These are:

1. Concept;
2. 35% (schematic or sketch) design, the most common;
3. Production drawings;
4. Construction.

Survey respondents also provided a wide range of views on either when they had encountered VE studies or when they thought VE studies should be undertaken. Again, the most common target point was seen to be up to 35% design, although there was fairly strong support for studies up to 60% design or just into working drawings.

When these views are taken in the context of North American construction culture there is a strong argument for suggesting that the most common target point of 35% provides an optimum locus for a study since costing data is more readily available in the form of the cost estimate and savings can easily be identified, thus justifying the bottom line figure of VE fee versus savings specified. In addition, any changes to design are more easily introduced at this point than any stage beyond sketch design.

7.4.   THE 40 HOUR WORKSHOP

The 40 h workshop is the most commonly adopted approach for undertaking VE studies and, as indicated above, the role that government has
had in supporting VE development cannot be ignored within this
context.

The authors have reservations about the usefulness of the 40 h workshop as the primary means of undertaking a VE study, especially at 35%
design. There is a considerable amount of information to assimilate in
that time period, in terms of overall project concept, specification and
cost estimate. In addition, group dynamics are at work and unless the
VM team are used to working together it will take some time before the
team is cohesive enough to allow the creative component of group
dynamics to flow freely. The experience of the authors when undertaking a VE training workshop was that the requirements of information
assimilation and the need to show identifiable savings is likely to force a
VE team down the cost reduction/substitution route rather than get to
grips with functional analysis, which does take considerable time.
Furthermore, VE techniques focusing attention only on the high cost
elements makes the assumption that the project concept is correct. This
can be an arguable assumption and one of the case studies in Chapter 4
demonstrates clearly that major savings can be made through substantial
concept revisions.

The survey data reveals a number of distinct views on the 40 h workshop. There was a broad consensus that on small to average sized projects
40 h was a sufficient time period for project assimilation. On more complex projects, especially where there was a high degree of mechanical
and electrical engineering work, figures of between 80–120 h were
reported. One respondent did, however, comment that 40 h was too little
but 80 h too great. Another respondent indicated their company used a 56
h workshop broken down as follows:

1.   Three days intensive study followed by;
2.   A one week break and a further two days of study followed by;
3.   A one week break and a final two days of study.

This particular firm had opted, therefore, for a seven-day study spread
across three to four weeks. The authors also interviewed VE practitioners
in Chicago who had a preference for a similar type of VE study. Their
argument for conducting the study spread across a number of weeks was
the easing of disruptions to office routines as well as allowing the subconscious mental processes of the VE team to gestate on the problem, a
key issue in stimulating creativity.

The benefits and disadvantages of the 40 hour workshop

These can be broken down into three distinct areas: (i) the benefits of an independent review; (ii) the problems associated with a review being undertaken on an existing design team; (iii) the disruption to the process.

The benefits of an independent review

The overriding benefit of the independent 40 h workshop was seen to be the objectivity that a new team and a fresh outlook could provide on a project. Furthermore, provided a team was selected on the basis of the appropriate mix of disciplines and expertise it could provide the most up-to-date technology for an additional design review at least cost. Other positive comments included:

• Proving that the initial design was the most effective;
• Lowering cost. Some respondents suggested a cost saving representing a factor of five compared to VE fee could be expected and if all the VE proposals were implemented this could increase to a factor of 20–30 times the VE fee;
• Improving the project schedule;
• Cost and quality improvements for the client and contractor;
• Improving value;
• Improving the technical specification;
• Improving the management of a project;
• Peace of mind for the client.

For the sake of balance, however, it must also be pointed out that a number of respondents saw little benefit at all for a VE study.

The problem of the existing design team

It cannot be denied that unless handled with extreme diplomacy and care, the independent 40 h workshop is perceived as being adversative by existing design team members. Common comments made by design team members were:

• There was no objection to a study provided the VE team comprised of competent professionals;
• The design team had already thought of many of the ideas that were generated by the VE team and subsequently rejected them;
• The VE team were not operating under the same constraints as the design team;

- Pride of authorship of the existing design;
- Yet another hoop that the design team have to pass through;
- The existing team are very familiar with the project and the outsiders provide yet another approach to cheapening a project, sacrificing quality and aesthetics;
- VE studies can easily reflect negatively on the original designers.

One comment from a architectural delegate on the training workshop undertaken by the authors and quoted in full, provides a clear insight into how the 40 h workshop could easily be seen by design team members:

'The 40 hour workshop precludes an integrated approach. It sets itself up as 'gunslingers coming into town to clean up a mess'. It sets a bad attitude, forces participants to go through the exercise without creative thinking or contemplation. The big VM firms can have a cache of apparently novel solutions to the one-off procurer of VE. Doing it day-in and day-out would reduce creativity'

The next section highlights the effects on the design and construction process itself.

7.5   THE DISRUPTION TO THE PROCESS

The overwhelming majority of comments from survey respondents relate to the impact of a VE study on the total process. The main comments are as follows:

- It is time consuming, especially on small projects. The interruption to the flow of work for the design team can be more costly than the savings made;
- There is considerable time wasted by the design team in reviewing VM proposals, writing responses and satisfying themselves of a different design than they had specified originally;
- The changes suggested were unnecessary;
- The time for re-design was not allowed for in the design schedule.

A number of other comments related to the impact on project specifications, the subject of the next section.

7.6   THE IMPACT ON PROJECT SPECIFICATION

Two clear comments stand out from respondents about VE proposals and the study process in general. The first is very much concerned with a lowering of quality of a project due to the pressure to save money. The second, in part a function of the 40 h workshop itself, related to the fact that in such a short space of time the VE team could not hope to fully understand the project in comparison to the existing design team. Any changes made may or may not be in line with the integrity of the total project concept.

7.7   DESIGN LIABILITY

The design liability for VM proposals was a thorny issue which the authors have yet to resolve. The basic problem is two fold: (i) the design team should take all design responsibility for VM recommendations implemented, since they are only to be viewed as recommendations and it is the design team that can either accept or reject them; (ii) the VM team should take design liability for any recommendations implemented, since as competent qualified professionals, they have an ethical responsibility the moment they review a design.

VM practitioners tend to take the first view. One VM practitioner, a qualified and practicing engineer, took the view suggested in (ii) above based on an extensive review of the legal situation in North America. However, this engineer was referring to the situation where he had prepared drawings as a part of a VE study. The survey responses, from contractors, architects and engineers, indicated an agreement with (i). Presumably on the basis that the moment an idea is incorporated into a design by the design team they have vetted that idea, found it to be sound and therefore its implementation has become their responsibility. The authors are aware of no US case law on this subject.

The US Navy have an interesting approach based on their procuring power as a client. In a situation where a design team having vetted a VM proposal refuse to accept liability for inclusion into the design and continue to reject it, then in this situation the Navy are prepared to accept liability as client on behalf of the designers.

7.8   THE VE CLAUSE IN DESIGN TEAM CONDITIONS OF
ENGAGEMENT AND REMUNERATION FOR CHANGES

The inclusion of a clause in the conditions of engagement for design consultants that a VE study would be undertaken appeared to ease the passage of the study and the subsequent feelings of individuals about the outcome. There were differing views expressed on the remuneration of consultants for any changes that were included in the design and stemming from VE proposals. The majority view appeared to be that no additional remuneration was made. However, there was evidence to suggest that additional fees were paid to some consultants.

7.9   THE USE OF FUNCTIONAL ANALYSIS

The use of functional analysis in value engineering is part of the root and branch qualification procedures for a certified value specialist. The empirical evidence suggests that it is not a major component of North American VE studies in construction. There were some VE consultants that used functional analysis in all studies they undertook and they argued quite forcefully that any study that did not incorporate functional analysis as an integral part of the project appraisal was not value engineering.

7.10 SUMMARY

The message of this chapter is that the standard North American approach to value engineering has to be viewed in the context of the cost control procedures that operate in that construction culture. The influence of the government in the development of VE in construction also cannot be dismissed. There are numerous problems with the 40 h workshop approach to VE. The 40 h workshop has its benefits, primarily, objectivity and technology transfer but if it is to be utilized in a British context account has to be taken of the adversative problems, the disruption to the design and construction process and the assembly of an independent creative VE team. The size and complexity of a project, the target point for a study and the express choice of using functional analysis also has to be taken into account. On larger projects with high levels of building services a longer period for the study may be required.

CHAPTER 8 ⎯⎯⎯⎯⎯⎯⎯⎯⎯⎯⎯⎯

PROJECT ECONOMICS AND THE CLIENT VALUE SYSTEM

⎯⎯⎯⎯⎯⎯⎯⎯⎯⎯⎯⎯⎯⎯⎯⎯⎯⎯⎯⎯⎯⎯⎯⎯⎯⎯

8.1   INTRODUCTION

This chapter is concerned with defining and distinguishing between a cost and value management service. It also attempts to establish the analytical processes that have to be established if the 'client value system' is to be made explicit for project development.

In brief, value management is defined as a service that takes place at the front end of a project where the primary emphasis is on making explicit the client's value system through the use of functional analysis and other problem-solving tools. Cost management is defined as a service where the primary emphasis is on cost reduction or substitution. It encompasses much of what the quantity surveyor currently undertakes although it can be extended by using a multi-disciplinary design team to focus on cost savings. An economic management service for projects involves the combined use of both value and cost management services.

A brief overview of projects and project management is also provided. An introduction to systems thinking is given, an approach which argues for adopting a 'whole picture' view of events rather than focusing on detail. Projects as systems are discussed and project management is defined. An outline is also provided of where value management procedures can fit into the overall project management process and its interrelationships with the client value system.

8.2   VALUE MANAGEMENT AND COST MANAGEMENT

Kelly and Male (1991) argue that the total economic management of projects involves considering both cost and value. The former is viewed more as an objective characterization of client requirements expressed in

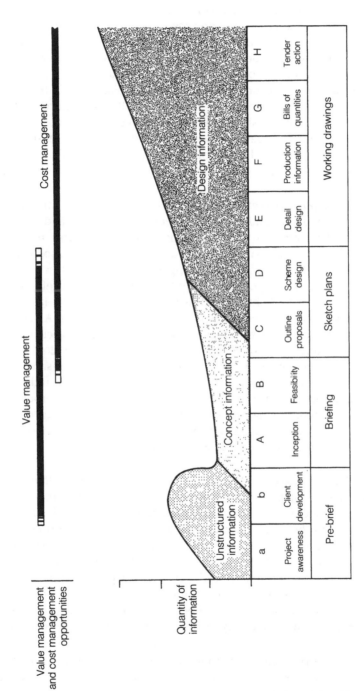

Figure 8.1 Value management and cost management opportunities.

monetary terms only. The latter, however, encompasses cost but also takes account of the subjective decision making criteria of the client organization in perceiving what is or is not an acceptable level of cost for any given level of project performance and technical specification. Hence, it is important to distinguish between **value management** and **cost management**.

Value management occurs much earlier in a project's life-cycle whilst cost management occurs in the later stages (Figure 8.1). In addition, the authors contend that there are particular skills attached to both types of service. The more easily identifiable services within cost management involves traditional quantity surveying services such as feasibility studies, cost planning, the production of bills of quantities, tender evaluation and on-site cost measurement for the contractor.

(a)   Value management is defined here as service that utilizes structured functional analysis and other problem solving tools and techniques in order to determine explicitly a client's needs and wants related to both cost and worth. The cost aspects of the service can be ascertained by the use of traditional quantity surveying services. The assessment of worth and its relationship to cost can only be ascertained with reference back to the views of the client or end-user. Hence, value and its related term worth encompass subjective aspects associated with cost. A value management service attempts to integrate, therefore, client subjective and objective cost criteria for project analysis and evaluation. Value management services involve considerably more emphasis on problem solving, in-depth functional analysis, the relationship between function and cost, and a broader appreciation of the linkages between a client's corporate strategy and the strategic management of the project. Value management is more involved, therefore, in the acceptability to the client of the cost implications associated with changes to the project concept.

(b)   Cost management is defined here as a service that synthesizes traditional quantity surveying skills with structured cost reduction or substitution procedures using a multi-disciplinary team. In this latter instance it would involve the generation of ideas by using group creativity techniques to optimise the insight and expertise of the multi-disciplinary team. Kelly and Male (1991) considered that cost management skills should be viewed more in the realm of objective client project appraisal since they may be undertaken without reference to client value judgements since they are already likely to be embedded in the design to date. Cost management does not make any major changes to the project concept.

It must be pointed out, with reference back to Figure 8.1, that there is a window of opportunity at or around 35% or sketch design where a choice exists to make either:

1. Substantial changes to the project concept since there is still fluidity in the design process and no major costs have been incurred except that proportion of the design budget spent to date, or;
2. The project concept can remain intact and a cost management service is opted for where the major revisions to the project are obtained through cost omission or substitution.

The next section explores in more detail how value management and cost management services can be integrated into a total project economics service.

8.3   A PROJECT ECONOMICS SERVICE

Through an analysis of the research data obtained from North American clients, projects and the questionnaire surveys undertaken by the authors, coupled with the existing cost control procedures in the UK, there are a number of alternative mechanisms that can be used for project auditing as part of an economic management service for clients. These alternatives for managing the economics of projects are:

1. Value management studies using functional analysis and other problem solving tools and a multi-disciplinary design/building team. This attacks directly the issue of understanding the client value system and attaching notional relative costings, especially in the early stages of a project, to function and worth. The client is in a situation of understanding the impact of the decisions made at the early stages on the project in its entirety with its ultimate functionality. The focus of this type of study is on ensuring that the client concept for the project is correct.
2. Structured cost management studies, utilizing the 'job plan' and using the criteria of cost reduction or substitution and a multi-disciplinary design/building team. The emphasis here is on utilizing the collective wisdom and experience of the multi-disciplinary team through group dynamics to suggest alternative materials or components as well make small revisions to the project concept, for example, re-configuring the spatial requirements of lift shafts in an office block.
3. Cost management studies using traditional quantity surveying skills.

Methods 1. and 2. have the advantage of using a multi-professional, structured, creative group approach whereas method 3. is undertaken within the context of a single profession. In addition, the changes that are suggested in 3. are very much dependent on the pro-activity of the quantity surveyor in design team meetings (Palmer 1992). In addition to these three approaches to auditing the economics of a project can also be added the choice of:

1. A team co-ordinator or facilitator chairing a project economics study with the existing design team or;
2. An independent project economics team is brought together under the chairmanship of a facilitator.

These can also be placed against the stages at which project audits can be undertaken, namely: (1) pre-brief; (2) immediate post brief/concept; (3) sketch design; (4) production drawings, and (5) construction.

This combination of project auditing mechanisms when read in conjunction with the stages of the project life-cycle that can be targeted and the tools and techniques that are available for project analysis provides an overall **project economics** service. In this respect it is useful to think in terms of these types of studies being conducted by a **facilitator** who is charged with the responsibility of deciding the appropriate auditing mechanism for a project in the context of its stage in the life-cycle and the particular problem to be addressed.

Having defined value and cost management within the broader service of *project economics* the next section outlines some key ideas in strategic management in the context of understanding the client value system.

8.4   THE CLIENT VALUE SYSTEM

Clients are organizational complexes that, as ongoing entities, are involved in an exchange relationship with a broader environment. Table 8.1 sets out different types of clients that can be encountered in the construction industry (UCERG 1975).

Table 8.1   Client types

Client types	
Public sector	National public Local public Public corporations
Private sector	Individual domestic Individual commercial
	Corporate commercial Corporate industrial Corporate developer

Private sector individual clients (domestic and commercial) can be excluded from the discussions that follow since, as the categorization implies they represent the very small client, perhaps the home owner/

house purchaser or sole business proprietor. In addition, the privatization programme of the Conservative government may well have minimized the importance of certain types of public sector client types from the Table as they will now be found within the private sector types.

The senior management team within the client organization can be viewed as steering that organization through an external environment over-time using either an explicit or implicit strategic management process. Therefore, the strategic management process within the client is concerned with managing the long term relationship of the organization with the environment within which it operates. Central to the strategic management process is (Male 1991):

- A future orientation;
- An ability on the part of managers to make decisions about the on-going relationship of an organization with its environment;
- The management of different types of change that the organization will face.

Organizations also involve the interaction between people and a technical system, perhaps machines or computers. Therefore, any organization will involve both *structure* – physical layout, hierarchical power relationships, reporting relationships and the pattern of formal and informal communications – and *process* – the on-going activities of the organization involved in converting inputs into outputs, monitoring, controlling and decision-taking (Wilson 1990).

From a corporate perspective the way in which organizational structure and process are configured and adapted through the strategic management process creates a 'value chain' (Porter 1985). The value chain comprises important key activities that are undertaken within the organization and involve transforming inputs into outputs. This transformation process creates value for the ultimate buyer. For example, a car manufacturer purchases raw materials, labour, and components, etc., which are used in the production process to produce cars. The manner in which the car manufacturer has organized the physical layout of the car plant in terms of machinery, the interrelationships between different processes in making cars, the manner in which management, labour and machinery interact, together with the way inputs are procured and output – the finished cars – are priced and sold creates value. The key activities making up this process for the car manufacturer comprise the value chain and result in a profit for the company and provide value to the purchaser. The value chain is a product of (Porter 1985):

- An organization's history;
- Its strategic management process in terms of strategy formulation, choices made and implementation procedures used;

• The cost and resourcing implications for the organization of these key value activities,

The value chain has, therefore, an internal organizational component related to structure and process and an external environmental component related to suppliers and buyers.

The notion of the value chain is conceived from within business strategy where the idea of the 'market' is easily understood. However, a good case can also be made for considering that the ideas behind the value chain can be applied to 'social markets' that often exist in construction, for example, those projects that are required to meet a social need, hospital buildings being a case in point.

8.5   THE STRATEGY OF ORGANIZATIONS AND THE STRATEGY OF PROJECTS

The preceding section has outlined the fact that from a corporate perspective the manner in which organizational structure and process are configured through the strategic management process creates a value chain that is rooted in the transformation of inputs into outputs. Construction projects are triggered from within client organizations. These triggering events will evolve from a need created through the interplay between external environmental changes and organizational structure and process.

The origin of construction projects are intimately connected with the evolving realization of a client's corporate strategy seen in the context of some form of market need – be it business or social – and the requirement to continue that strategy through the creation or modification of structures designed and built by the industry. When placed within this framework the industry has to understand and respond to a diversity of project origins, client types and hence differing client value systems.

The triggering event for the construction project, as a response to a market need, has to be understood within the context of the client value system and the associated strategic management process. Therefore, the design and construction process is also intimately connected with and forms part of the client's value chain (Reve 1990), with the newly built or modified structure intended to hopefully enhance and not detract from the ongoing client's corporate strategy.

At project inception there is a close linkage between a client's corporate strategy and the strategic management requirements of a project. It is precisely in these early formative stages of the project that insufficient attention is given to clearly identifying project objectives, their interactions and grappling with the resolving the ambiguities and uncertainties that are inherent in a project. It is this lack of time and effort at project inception that can contribute to its eventual success or failure (Morris and Hough 1987).

At the project awareness stage, information will be unstructured (Figure 8.1). It is likely that this will be expressed in terms of a broad project requirement, for example, the requirement for a new school, laboratory etc. Project awareness, through the triggering event, will occur deep within the client organization and economic, organizational and internal (as well as external) political factors will have a direct impact on the development of initial project requirements (Churns and Bryant 1984). It is these very influences, operating through the client value system, that should be encapsulated within project requirements expressed in the design brief. These influences may also lead to a project brief that represents a 'wish list' of all the interested parties that have a vested interest in the project (Kelly and Male 1987).

8.6   SYSTEMS THEORY AND ANALYSIS FOR THE MANAGEMENT OF PROJECTS

The essence of systems analysis is to look at the world in terms of 'wholes' and attempts to determine the properties of 'wholeness'. By adopting a systems perspective thinking is holistic, taking a broad macro view. Systems thinking and analysis adopts a perspective, therefore, that views the world in a way which says the effect of the whole is greater than the sum of the individual parts that comprise it. As an example, the word 'orange' conjures up in a person's mind a picture of a spherical object with a thick dark yellow surface. It is only when one delves deeper behind this higher level notion of orange that the ideas are encountered of an outer skin, the pith, juice etc. These ideas also form separate concepts or 'wholes' in their own right but they all form part of the higher level concept 'orange'.

Checkland (1981) sees systems thinking as being concerned with organized complexity. A human activity system, such as a construction project and also representing organized complexity, can also be viewed as a complex grouping of human beings and machines. Therefore, total system characteristics will differ qualitatively and quantitatively from those of the constituents (De Greene 1970, 1981). For example, with reference to a nuclear power station, the effects of the total concept will be quantitatively and qualitatively different than what goes on during design and construction – which can be considered as sub-systems of the project system, the nuclear power station.

Systems thinking and analysis is attempting, therefore, to investigate and understand phenomena at the conceptual level. In a construction context, the project concept of 'nuclear power station' can conjure up emotive issues of nuclear waste, the Not In My Back Yard (NIMBY) protest, as well as practical issues such as a cheap energy source. In this instance the total concept view of the nuclear power station is investigated through systems analysis.

Systems are embedded in environments

A system exists within an environment and this is a function of the hierarchical level under analysis. There is a reciprocal relationship between the system and it's environment – the system will be modified by the environment and modify it in return. Feedback mechanisms operate, therefore, between systems and sub-systems and creates a set of interdependent relationships that have to be understood. For example, cost planning on a construction project is at one and the same time a feed back mechanism to the client and design team and is also a control. Taking the nuclear power station as an additional example, the environment of the total project will be different than that faced at say the design stage.

In using systems analysis it becomes important to ascertain the boundary between the system and the environment (De Greene 1970, 1981). This will determine what is within the analysis and what is not. However, one of the main problems of using systems analysis is that where the boundary is set to delineate one system from another is very much up to the analyst.

Systems as hierarchies

Systems form a hierarchy (Jenkins 1969, 1981) and a system, as an organized complexity, is arranged hierarchically in levels. Each level has properties that do not exist at lower levels, termed emergent properties, and the properties possessed by each level are meaningless to those lower down (Checkland 1981). Going back to the example of of the 'orange', the total system concept is the spherical object with the deep yellow outer surface and this will have its own properties in the way it interacts with the surrounding environment. The next analytical level down, the skin, will have its own properties – it is hard, difficult to puncture, is pitted and tastes bitter.

Each level in the hierarchy can be considered to act as a constraint or control on the next level down. Figure 8.2 sets out the nuclear power station example as a series of systems and sub-systems arranged in a hierarchy. The effects of the total project concept level – level 1 have

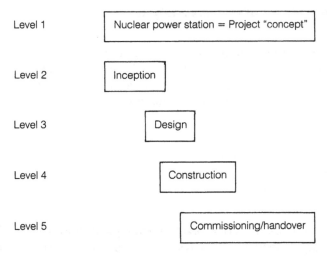

Figure 8.2   The project as a systems hierarchy.

already been discussed above. Level 2 – the inception/feasibility sub-system will involve the client organization determining project objectives, undertaking feasibility costings, pulling together internal teams. At level 3 – the design sub-system – design teams will be brought together and design objectives and time scales allocated. Therefore, there may be issues operating at level 1 (total system level) that will act as a control or constraint on level 2 (inception/feasibility). An example could be community action necessitating the siting of the project which may have an effect on cost and thus feasibility.

Using systems thinking to analyse the project in this way allows the decision-maker to explore the different effects of each level in the project and the kinds of issues that will be involved and that are only relevant in that phase or level in the hierarchy. However, it also allows the broader picture to be re-assembled as knock on effects are uncovered between each stage – interdependency.

Systems have objectives

System have objectives and these will often be in conflict. Compromise will be essential to obtain the best overall result (Jenkins 1969, 1981). In using systems thinking as an approach to project analysis it becomes imperative, therefore, to uncover the objectives that are operating at each level in the hierarchy, to determine if any are in conflict and thus what is the impact of this conflict on the project.

The building blocks of systems analysis

To summarize, systems thinking is holistic and asks of the analyst and decision-maker to view the world in 'wholes'. A number of systems concepts have been introduced and explored in the context of examples. The basic building blocks of systems thinking are:

1. Environments;
2. Boundaries;
3. Systems and sub-systems;
4. Hierarchy;
5. Control and feedback;
6. Interdependency,

Critical path networking is one practical example of how systems thinking is used. Perhaps implicitly, to analyse a construction project. The next section explores a project as a system.

8.7   PROJECT SYSTEMS

What is a project?

Borjeson (1976) indicates that a project has a temporary status and is started, implemented, evaluated and wound-up as a separate entity. In addition, Borjeson indicates that projects have two important characteristics: they are organizations for **learning** and for **change**. A project can be defined therefore as:

'a temporary activity with defined goals and resources of its own, delimited from but highly dependent on the regular activity' (Borjeson 1976:11).

Morris and Hough (1987:3) define a project as:

'an undertaking to achieve a specified objective, defined usually in terms of technical performance, budget and schedule'

Project objectives or goals are achieved through people who are arranged in a project organization which can be defined as:

'a temporary organization, group or groups mandated with the task of carrying out a project' (Borjeson 1976:11).

Common themes in these definitions are finite tasks or activities, objectives, money, performance, people and groups. These comprise the 'grist' of the management of projects. Therefore, from a construction perspective, a project is a process that has to be managed and starts in the client organization, can take years to evolve and finishes either with owner-occupation, the leasing, renting or buying of a facility. A project starts and ends with a client or end-user 'organization'.

Problems with projects

Communication requirements in complex projects are overwhelming in comparison to more traditional manufacturing processes and there is a requirement for a great deal of interaction and negotiation. Furthermore, as the project proceeds from inception through to completion and handover – the project life-cycle – there will be a different mix of managerial, professional and technical people (Sayles and Chandler, 1971:8). For construction projects, as one proceeds through the project life-cycle there will also be different numbers of organizations involved. Once a project is in the on-site construction phase, with the increasing use of sub-contracting the number of organization involved blossoms almost exponentially. For example, Morris and Hough (1987) indicate that on the Thames Barrier project there were 450 sub-contractors involved at peak of construction activity.

Projects are problem-orientated, dynamic and will usually require inter-disciplinary effort. However, the integration of different specialisms requires a reorientation of the professional to the project whilst allowing the individual to maintain his ties with professional groupings within his field of expertise (Sayles and Chandler, 1971).

'Major projects' are defined by Morris and Hough (1987) as those that are particularly demanding due to their: (a) size; (b) complexity; (c) schedule urgency; and (d) demands placed on existing resources or know-how.

Morris and Hough also contend that the term 'major' is a relative one and does not necessarily always involve large sums of money. There can equally be small value, but highly complex and difficult projects.

Sayles and Chandler (1971) point out that complex projects highlight the tension between the needs of the project, constituent organizational needs, narrow disciplinary approaches and the problem orientation of the project. In addition, project teams are temporary in nature whereas participants to a construction design team, for example, are rooted in more permanent organizational and social groupings. Therefore, joint endeavours, such as building projects, require a different style of management. In essence the project manager is dealing with the management of both inter- and intra-organizational relationships. Inter-organizational relationships involve, for example, the fact that design team members are in reality representatives from other organizations. Intra-organizational relationships will involve those within the design team or project organization itself or perhaps within the project manager's own employing organization.

Project planning

Planning is not synonymous with forecasting but is a dynamic process by which inside (perhaps the client and in-house project manager) and outside interests (the other consultants) arrive at a new balance of power to set a structure for decision making on a project (Sayles and Chandler, 1971). Put simply, project planning sets up power relationships between individuals, teams and between organizations.

Project planning requires a clear division of the project into stages. The two main motives for this are **manageability** and **controlling possibilities** (Borjeson 1976). In this latter instance the main thrust is anticipating, making allowance for or attempting to control the future.

The planning of projects requires, therefore, an orientation that is able to anticipate, manage and control the dynamic interactions of future possibilities as they become reality. The manager of projects requires skills that are orientated to the future and present; skills of strategy and implementation.

The skills of the project manager

The manager of projects has to concentrate on getting work done through others. This will require different skills and a different 'theory of management' than for traditional line management. Furthermore, the project manager is primarily dealing with rates of time and organizational processes, not technical variables (Sayles and Chandler 1971). Based on research data, Sayles and Chandler indicate that the behaviour of project managers falls into the categories of: (1) bargaining; (2) coaching or cajoling; (3) confrontation; (4) intervention; and (5) order giving.

These are essentially wrapped up in the interpersonal skills of negotiation and persuasion and reflect the increasing use of power, from covert to overt. In addition, Morris and Hough (1987) suggest that the skills of the project manager are fundamentally concerned with progressing the project through its life-cycle and will encompass: (i) leadership; (ii) organization; (iii) financing; (iv) planning and control; and (v) the contracting of third parties. The next section discusses project management.

8.8   PROJECT MANAGEMENT OR THE MANAGEMENT OF PROJECTS

The Chartered Institute of Building (1982) defines project management as:

'The overall planning, control and co-ordination of a project from inception to completion aimed at meeting a client's requirements and ensuring completion on time, within cost and to the required quality standards.'

Anthony Walker (1989) gives a more elaborate definition:

'The planning, control and co-ordination of a project from conception to completion (including commissioning) on behalf of a client. It is concerned with the identification of the client's objectives in terms of utility, function, quality, time, cost and the establishment of relationships between resources. The integration, monitoring and control of the contributors to the project and their output, and the evaluation and selection of alternatives in pursuit of the client's satisfaction with the project outcome are fundamental aspects of construction project management.'

To the authors, the essence of managing construction projects is concerned with controlling time, cost and quality within the context of project functionality. Figure 8.3 sets out these ideas within the framework that will be used subsequently to discuss value management in the context of a broader project management orientation.

Construction projects, and their management, are intimately linked with the client's corporate strategy and the on-going creation of value for the

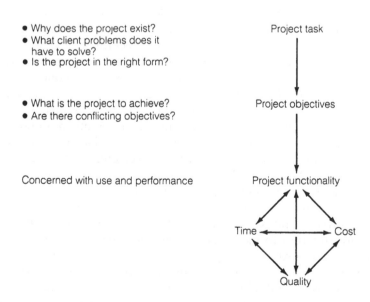

Figure 8.3   The context of projects.

client. The creation of value for the client is intertwined with the exploration and resolution of project functionality. Value management is a philosophy and set of tools and procedures that analyses function.

8.9   VALUE MANAGEMENT AS A STRATEGIC PROJECT MANAGEMENT TECHNIQUE

When taken in the context of the foregoing analysis, value management tools and procedures can be seen as part of the early strategic management inputs in developing the project concept and design brief. Therefore, through the use of functional analysis, part of the value management process in the early formative stages of a project would be to identify clearly and explicitly the **project task** – the reason or reasons for the project's existence in the first place – by identifying the clients: **Needs**, that is, the primary functions of the project and what it is to achieve, and, **wants** or embellishments.

It is the latter area where much of the unnecessary cost in a project is buried but can also add value provided the client is made aware of the impact of these wants on project costs.

8.10 SUMMARY

This chapter has argued that the economic management of projects – a project's economics service involves the integration of value and cost management services. Events and processes occurring deep within the client organization act as the stimulus for construction projects. These events are rooted in and should be interpreted in the context of the client's value system. This is the way in which key strategic activities are configured into a value chain, representing the choices that the client organization has made historically in transforming inputs into outputs to create value. There is a close linkage between the strategic management of a construction project and its associated implementation through design and construction, its contribution that it will eventually make to the value chain, and the ongoing client corporate strategy.

The driving forces behind a project involve issues of organizational, political, economic and psychological behaviour. These forces have to be understood and transformed into parameters defining a construction project. In this context, providing an economic management service for a client has to entail a capability to uncover these forces, make them explicit, attach costs and allow a client to re-interpret these costs within a value framework. Ensuring that representatives from the key value activities are involved in the deliberation of the client project team would appear vital in order to give the project its correct impetus and momentum from the outset.

The use of functional analysis and other problem-solving tools within a multi-disciplinary team can provide insights into projects for client organizations. The primary focus of value management, by using these procedures, is in assessing the relationship between function, cost and worth. This chapter has argued that it is essential to tap explicitly into a client's value judgements about a project and encapsulate these insights into the decision-making process when developing the design brief.

This chapter has argued that the major impact of value management is in the earlier stages of a project, when there is a close linkage between the corporate strategy of the client and the strategic management of the client's construction project. However, it is at these stages that it becomes more difficult to provide a bottom line figure of cost savings on a project. Value management procedures provide an excellent method for integrating the design/building team, the client and end-user.

The early formative stages of a construction project are involved therefore in 'strategic project management' which involves:

(a) making a number of major decisions involving cost and value issues;
(b) handling and understanding the creation of information that expands exponentially as the number of individuals contributing to the design process increases.

Kelly and Male (1991) concluded in their study that the integration of cost management and value management skills into a unified project economics service to the client can best be instigated through project management. The cost of such a service is likely to be accepted by the client where it is incorporated into the fee for a broader, more integrated strategically orientated project management service. This chapter has also argued that a separate project economics service is also a viable alternative.

Project management is defined usually in terms of controlling time, cost and quality. By adding the further dimension of **project functionality** and analysing this within the client constellation intimately involves a value management team with strategic management and the client value system. Understanding value activities (or the value chain) would provide considerable insight into the determinants of project functionality for the client and ensuring that the correct in-house team is assembled for the project and subsequently incorporated into any value management study. Thus, the best vehicle for utilizing value management as part of a strategic project management service that incorporates the total economic management of a project would be through analysing and controlling the parameters set out in Figure 8.4.

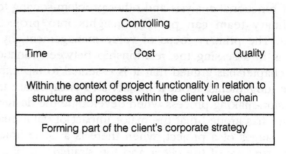

Controlling		
Time	Cost	Quality
Within the context of project functionality in relation to structure and process within the client value chain		
Forming part of the client's corporate strategy		

Figure 8.4   The relationship between value management and project parameters.

This concludes Part Three. In Part Four the key tools and techniques of a project economics service will be drawn together into a proposed UK method.

PART FOUR

A PROPOSAL FOR A UK IMPLEMENTATION OF VALUE MANAGEMENT

CHAPTER 9 ⎯⎯⎯⎯⎯⎯⎯⎯

FUNCTIONAL ANALYSIS METHOD

9.1 INTRODUCTION

The analysis of 'function' in the design of buildings has been considered important for at least two thousand years. Virtruvius (1st century BC) devotes considerable energy to the proper description of spaces within a dwelling, giving consideration to their function, their position and orientation and their purpose as required by house owners of differing professions or businesses.

The analytical description of function is more recent and has two apparently unrelated sources. The first can be traced to the operational research technique of objectives hierarchies and the second to functional analysis diagramming developed within the framework of value management.

Functional analysis diagramming is attributed to the developing techniques of value management within the manufacturing industry of the US in the latter part of this century. The techniques have been transferred to the construction industry with mixed success. However, a re-examination of the basic philosophies gives clues on the application of new techniques to the briefing and sketch-design stages of construction projects.

In this chapter the words 'component' and 'element' are given specific meanings. A 'component' is defined as a single manufactured item. For example; a brick, a hinge, a gear wheel, a length of pipe, a switch, or as a unit comprised of a number of components. For example, a window, a roof truss, a wash basin, and a kitchen unit. An 'element' is defined here as in BCIS as a unit of construction 'that fulfils a specific function or functions irrespective of its design'. For example, a window, and an external wall, etc.

9.2   FUNCTIONAL ANALYSIS PRACTICE IN MANUFACTURING AND CONSTRUCTION

Although functional analysis has remained the cornerstone of value management for almost fifty years many US practitioners find its structured use of little value in the performance of a construction-orientated value management exercise. The reasons for this can be explained through an examination of the characteristics of the two industries.

Manufacturing industry is concerned with the manufacture of components and the assembly of products. All products can be broken down into components. All products and components perform a function. For example, consider a hand-operated drill, (Figure 9.1). The function of the drill is to produce a hole which is carried out by rotating a drill bit. The hand drill has therefore a facility for holding it in the hand, a means of rotating, by hand, the drill bit and a means of securely holding the drill bit, the chuck.

In a functional analysis the function of each component is examined by asking the question 'what does it do?'. For example, in the hand-operated drill the large gear wheel on the side of the drill transfers hand-generated torque to the shaft. In a value management exercise the next question would be 'how else can this be achieved?'. A brainstorming exercise is held and other technical solutions generated. For example, pump action (mechanical, air, hydraulic), and ratchet action, etc. These ideas may be refined. For example, a pump action by a foot pump, leaving both hands to guide the tool.

However the problem is solved, one major fact should not be ignored, and that is the scale of the savings or sales opportunities which may flow from the idea. If a small tool manufacturer is producing one million hand drills per year and a value management exercise saves £1 on the manufacturing costs and/or creates a marketing opportunity which generates £1 per drill sold then that company gains £1 million per year. The costing of the alternative ideas may be complex but will involve the analysis of familiar materials and production methods. The process is therefore attractive to manufacturing because it highlights an effective means of

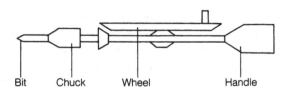

Bit   Chuck   Wheel   Handle

Figure 9.1   A hand-held, hand-powered drill.

generating alternative technical solutions relevant to components and products which, in the main, have large production runs.

There is another highly relevant feature of manufacturing-orientated functional analysis which is that the component is designed and manufactured by the client organization. The acceptance or rejection of ideas flowing from the functional analysis will be taken on by those of the client organization partaking in the exercise.

Construction, on the other hand, is concerned with the provision of accommodation (building) or a community utility (civil engineering). The function of a building is to provide an environmentally-controlled space suitable for the activity to be carried out within that space. The design of the building is a technical solution to the functional requirements of the space. Herein lies one major difference between the manufacturing industry and the building industry; manufacturing providing products and building providing environmentally-controlled space.

The other major difference lies in the scale of production. Manufacturing, with the exception of industries such as aircraft and shipbuilding, provides a large number of identical products for sale. Building, with perhaps the limited exception of housing, provides single products to order. Generally a £1 saving on a building or a £1 gain through a marketing opportunity is of no consequence.

Functional analysis within building, to be worthwhile, must therefore deal with functions of a higher order or more significance and unlike manufacturing must precede even the production of a prototype.

9.3   FUNCTIONAL ANALYSIS RULES

To head this section as rules is perhaps a little presumptuous as neither operational research texts in the case of objectives hierarchies nor value management texts in the case of functional analysis describe a fixed set of operations or rules. However, the perceived wisdom in these texts is summarized and collated into a working model.

(a)   Verb/noun definition most texts recommend that the function of an item or a system be expressed in as concise a phrase as possible, ideally one comprising just a verb followed by a noun. This discipline, although not absolutely essential and perhaps constraining functional exploration, allows an exact statement of the function which is readily understood. Where an item or system performs more than one function, all functions are listed in groups of verb plus noun. For example, the function of the main shaft in the hand held wheel brace illustrated in Figure 9.1 is to transmit torque; support handle; support chuck. Certain verbs such as

'provide' or 'allow' are not helpful and their use is discouraged. A list of some useful active verbs is given below:

amplify	establish	modulate
attract	filter	prevent
change	hold	protect
collect	impede	rectify
conduct	improve	repel
control	increase	shield
create	induce	support
emit	insulate	transmit
enclose	interrupt	

(b)   Functional definition/Technical solution. A technical solution to a problem is represented by a component or element. For example, where light is required in a room a technical solution is to install an electric light bulb. It is not possible to search for alternatives to a technical solution without first realizing the functional definition. For example, light is required in a room – functional definition – install a component which emits or transmits light. Technical solutions to this functional definition are, for example to install an electric light, install a sheet of transparent material between the room and a light source, install a gas light, etc.

A functional definition is in the vast majority of cases realized by first seeking a technical solution and then defining the functional performance of that solution. Research has demonstrated that creativity prompted by a functional requirement is satisfied by a technical solution and not another sub-set of a functional definition.

The recommended iterative process is therefore, seek a technical solution, then define the technical solution in terms of one or more functional definitions, then seek alternative technical solutions to the functional definitions.

(c)   Primary/secondary functions. Primary functions are defined as those without which the project would fail or the task would not be accomplished. Secondary functions on the other hand are those which are a characteristic of the technical solution chosen for the primary function and are not required. Both primary and secondary functions should be sought in order to fully understand a problem.

For example, an electric filament bulb satisfies the primary function of emitting light but also has unwanted secondary functions of generating heat, induces glare, looks unattractive etc. These secondary functions can be solved through the technical solutions of ventilation/air conditioning and by a lamp-shade. These technical solutions can themselves be represented by a function of control waste heat and shield light source. Further technical solutions can now be generated for these functional definitions. This process continues until technical solutions only can be found. The process of containing the problem is discussed below.

In these exercises it is important to constantly refer back to the prime function. In many cases considerable sums of money are spent solving secondary functions, ie. solving problems which are characteristic of the technical solution to the primary function.

(d) Cost/worth. In Chapter 3 cost and worth were defined as follows: *Cost* is the price paid or to be paid. It is often said that one man's price is another man's cost. Cost can be divided and distributed among elements and to some extent, functions.

Worth is defined by North American value engineers as the least cost to perform the required function or the cost of the least cost functional equivalent. Therefore worth is a target figure which should if possible be bettered. If the cost is less than the worth value for money is obtained.

In the context of this chapter the definition of worth is expanded to enable a cost/worth appreciation. The main axiom in this context is that secondary functions have zero worth. This logically must be the case since by definition the project does not fail if a secondary function is omitted. The task is therefore to correctly identify the prime functions. For example, consider a door within a fire-check partition in a plant room area, and also consider a door within a fire-check partition but between an office and the foyer of a quality hotel.

The prime functions of the door within the plant room are to maintain access, and prevent the spread of fire. Within a plant room a closed door is often an obstruction and therefore to permit the prime function of maintaining access the door must be held open except in the case of fire. Therefore the prime function, maintain access, is worth only the amount which is necessary to form a hole in the wall. The function, prevent fire spreading, is worth the amount paid for the least expensive fire-check door and automatic door closers. The design solution cost can therefore be checked against the worth of these two functions. Through creativity it should be possible to derive alternative technical solutions which cost less than the worth.

The prime functions of the door in the hotel are:

- Establish access;
- Restrict access;
- Prevent the spread of fire;

- Contribute to aesthetic environment;
- Insulate office environment from foyer,

All functions, except 'contribute to aesthetic environment' and 'restrict access' can be provided for by the least expensive fire-check door and are therefore worth its cost. The difference in price between the least expensive fire-check door and one which has an acceptable aesthetic standard is the worth of the function 'contribute to aesthetic environment'. The function 'restrict access' could be accomplished by siting the door behind the reception counter and therefore the worth is zero. Again having established a worth marker the current design solution should be examined to determine its cost and also creativity should result in alternative technical solutions which cost less than the worth.

9.4   CONSTRUCTION ORIENTATED FUNCTIONAL ANALYSIS

As stated previously, the function of a building is to provide an environmentally-controlled space suitable for the activity to be carried out within that space. The building has a number of characteristics:

1. It is comprised of manufactured components;
2. The components are constructed to form elements of a building;
3. The configuration of the elements of a building form spaces which are conducive to the activity to be performed within the building;
4. The building represents a stage in the corporate strategy of the client organization and contributes to the capital value of that organization.

Each of the above characteristics represent a level at which a particular functional analysis approach is appropriate. The levels above are incremental but are in fact in the reverse order of their chronological development which is:

(a)   Task – Level 1 represents the first stage wherein the client organization perceives a problem. This problem may be realized through a study of efficiency, safety, markets, profitability, etc. Currently, if a client sees a building as a solution to this problem, a contact with the construction industry is most likely to be made. The construction industry representative is most likely to concur that a building is a solution to the problem and will advise the client on how best to proceed. The client at this point steps onto the building procurement 'moving walkway' and is virtually prevented from stepping off until the keys of the new building are handed over.

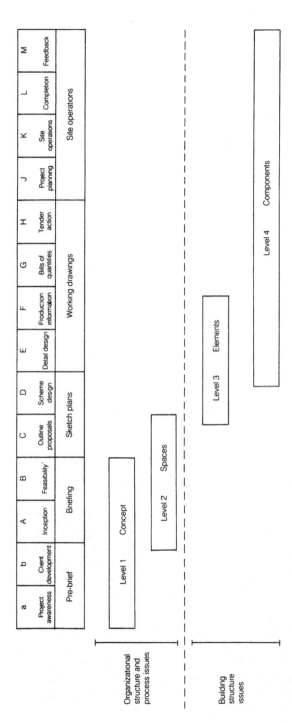

a	b	A	B	C	D	E	F	G	H	J	K	L	M
Project awareness	Client development	Inception	Feasibility	Outline proposals	Scheme design	Detail design	Production information	Bills of quantities	Tender action	Project planning	Site operations	Completion	Feedback
Pre-brief		Briefing		Sketch plans		Working drawings					Site operations		

Level 1   Concept

Level 2   Spaces

Level 3   Elements

Level 4   Components

Organizational structure and process issues

Building structure issues

Figure 9.2 The level of decision encountered in the project life cycle.

(b)   Spaces – Level 2 represents the stage where the architect or the whole design team are engaged in the preparation of the brief in conjunction with the client. Often a full performance specification of requirements is not available from the client and it is common for the design team to build-up a picture of the client's requirement for space interactively and iteratively through the production of sketches and cost plans.

(c)   Elements – Level 3 is the stage at which the building assumes a structural form. Often this stage can be incorporated into level 2 but is confusing from a functional viewpoint since the purpose of the element is to enclose and make comfortable the space. The element in no way contributes to the client requirement.

(d)   Components – Level 4 is the point where the elements take an identity in terms of built form. Contact with the client at this point is negligible since the client value system (discussed earlier) has been fully incorporated at previous levels. Components are selected to satisfy the requirements of the elements in terms of surrounding and servicing space.

The levels tend to be encountered at specific stages of the project life-cycle which are illustrated in Figure 9.2.

9.5   FUNCTIONAL ANALYSIS SYSTEM TECHNIQUE (FAST)

In writing on functional analysis Snodgrass and Kasi (1986) highlight the Functional Analysis System Technique diagramming method proposed by C W Bytheway in 1964. They present two types of FAST diagram which they entitle 'Technical FAST' and 'Task FAST'.

A Technical FAST diagram is one, such as that for the hand-drill (Figure 9.3) which tends to be linear and answers the question HOW? when reading from left to right and WHY? when reading from right to left. The How/Why logic is a useful concept and will be referred to again. Technical FAST tends to illustrate level 4 problems, ie., components.

Snodrass and Kasi give the impression that a technical FAST diagram is compiled by inserting the prime function on the left of the diagram and working across to the right, ie., in Figure 9.3 example, commencing at 'form hole' and arriving at 'turn wheel' by continually asking the question "how?". It is postulated here that this approach is incorrect and can be misleading. Where a technical solution exists, it is better to list all of the components of the particular technical solution, order them into related components and construct the diagram by working from right to left by asking the question 'why?'.

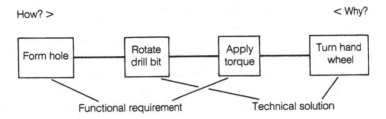

Figure 9.3 A FAST diagram of a hand-held, hand-powered drill.

For example, consider a domestic, gas-fired, low-pressure hot water heating system. The components are, gas-fired boiler, timeclock, room thermostat, hot water cylinder tank stat, pump, two-way valve, pipes, insulation, hot water cylinder, radiators, radiator valves.

A FAST diagram can be constructed by ordering the components at the extreme right of the diagram. A particularly useful method of doing this is to write a description of each component on an individual sticky yellow notelet which can be affixed to a desk, board, etc. The components can be grouped according to their function. Working from right to left produces

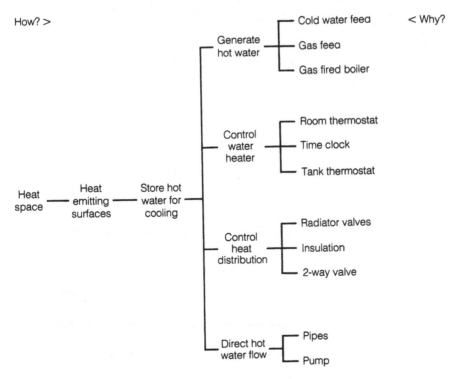

Figure 9.4 A technical FAST diagram of a low-pressure hot water heating system.

an hierarchy of functions, arriving ultimately at the prime function. This can then be checked back with the client to ascertain if this is the correct prime function required. In compiling the above diagram it was noted that the hot-water requirement could not be included since it was not related to heat space, the prime function. The diagram demonstrates that all functions are transitive ie., that all functions on the diagram can be related to 'heat space' irrespective of their position. The hot-water cylinder is a technical solution of the function 'store hot water for domestic use', which itself is a sub-function of 'generate hot water'. The fact that a gas-boiler can be used for this function is logically in tune with the function 'generate hot water' on the diagram but can not be included because of the transitivity rule. It is suggested here that a technical FAST diagram is compiled by listing all components and logically working from right to left to arrive at the prime function by asking the question 'why?'.

A task FAST diagram, unlike the technical FAST diagrams above, is described as having a primary function representing client need and four supporting functions representing client wants, namely: assure convenience, assure dependability, satisfy user and attract user. In a task FAST diagramming exercise a concept is taken and developed from left to right. There are very few illustrations of this type of diagram and those that are available relate to manufacturing. The example below has been developed for a fork lift truck of the type used in warehouses. The prime function is to handle loads in boxes or on pallets. Snodgrass and Kasi recommend the construction of the diagram by first entering the prime function on the left and working across to the right asking the question 'how?'.

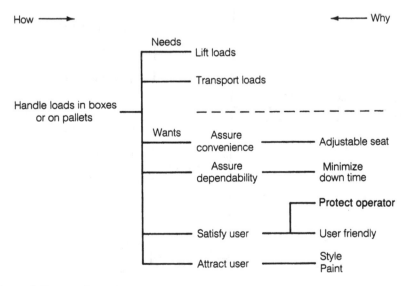

Figure 9.5   A task FAST diagram of a fork lift truck.

This diagram is not untypical of the style of diagram put forward as a concept diagram by value engineers. It has a number of flaws. The major flaw is that the 'needs' are an accurate functional definition of what is required whereas the 'wants' are related absolutely to the commonly perceived idea of a fork lift truck. Lift loads and transport loads could be carried-out by a remote control gantry crane and therefore an adjustable seat and protection for the operator is not required. While it is accepted that a task diagram is useful it is concluded here that the above form is not useful particularly when used for construction design.

9.6   DEVELOPMENTS IN FUNCTIONAL ANALYSIS TECHNIQUES

US construction based value engineers have taken the tools and techniques of manufacturing orientated value management and have applied them to the design stage in construction. This is arguably valid for level 4, the components of construction: bricks, blocks, pipes, electric wire, etc., since these are manufactured items. It is also arguably valid for level 3, elements of construction such as external walls, suspended floors, roofs, etc., since the techniques of functional analysis correspond to those for level 4. It is observed by Kelly and Male (1991) that the majority of US value engineering studies do concentrate on elements but tend to be driven by technical substitution based upon experience of an alternative solution rather than functional analysis.

There have been no significant developments in North American functional analysis methods since the work of Bytheway in the 1960s. This is unfortunate since it appears that the work of Bytheway preceded the work of the operational research community by at least eight years. However, references to objectives hierarchies and objectives trees are few in number and generally relate to specific examples and not to guidelines for their construction nor to axioms or rules.

Of the work described in this chapter the authors are confident of the approach in terms of levels. This subject has been rigorously examined and tested. The proposed method however has not been subjected to the same rigour although it has been applied to simple examples. Research work is continuing in this area.

9.7   CONCLUSION

In summary of the work to date it can be concluded:

1. Functional analysis is a powerful technique in the identification of the prime functional requirements of a project.

2. The subject has developed to the point where functions can be identified.
3. The levels at which functional analysis may take place during a project have been recognized.
4. The techniques for diagramming or other forms of representation unfortunately lies within the skills and experience of the value manager. To date no rules have been established. It should be noted that even within the related fields of objectives hierarchies, hierarchal-rules for diagramming do not appear to exist.
5. Value managers tend to currently work in levels 3 and 4 but the best opportunities appear to lie in levels 1 and 2.
6. The lack of cost models mean that new technical solutions are currently economically evaluated by a time-consuming costing of approximate quantities of resources. This approach must change with the formulation of cost modelling methods conducive to value management.

These conclusions demonstrate that considerable work has yet to be done particularly with regard to the establishing of rules of procedures for functional analysis diagramming.

CHAPTER 10   —————————

LIFE-CYCLE COSTING

—————————————————————————

10.1   INTRODUCTION

It is a fundamental principle of value management that the costs of all alternatives are expressed in life-cycle cost terms. This chapter is a brief outline of the theory and application of life-cycle costing. The chapter includes a discussion of the sources of data for use in life-cycle cost calculations as well as consideration of sensitivity and risk analysis.

10.2   DEFINITION

A definition of life-cycle costing is a technique for economic evaluation which accounts for all relevant costs during the investor's time horizon and adjusting for the time value of money.

10.3   RELEVANT COSTS

The costs which should be taken into account in a life-cycle cost calculation are:

1. Investment costs. These include:
    Site cost;
    Design fees (architect, qs, engineer etc);
    Legal fees;
    Building cost;
    Tax allowances: capital equipment allowance, capital gains, corporation tax etc;
    Development grants.
2. Energy costs. Heating, lighting, air conditioning, lifts etc.

3. Non-energy operation and maintenance costs. These include:
   Letting fees;
   Maintenance (cleaning & servicing);
   Repair (unplanned replacement of components);
   Caretaker;
   Security and doormen;
   Insurances
   Rates
4. Replacement of components (planned replacement at end of useful life).
5. Residual or terminal credits. In determining these credits in the context of a building it is necessary to separate the value of the building from the value of the land. Generally, land appreciates in value but buildings depreciate until they become either economically or structurally redundant.

Some costs are not relevant and are not accounted for in the calculation. These costs are either trivial in amount or do not affect the decision. In terms of the latter are those which are 'sunk', ie. the client has already expended money related to the project or 'unchanged', eg. if carrying out a project comparing double glazing with single glazing then need not take account of window cleaning costs. Similarly if rates or insurance are assessed on a M2 basis then exclude them from decisions unless comparing solutions of differing areas.

10.4   THE INVESTOR'S TIME HORIZON

The investor's time horizon is the period of time over which the investor has an interest in the building or sub-system. A client may require an exercise say to determine the cost effectiveness of the installation of a heat exchanger in an air conditioning system or may require an exercise to determine the cost effectiveness of a whole building.

The cost effectiveness of a building should be looked at in terms of the whole value of the building to the client. The effectiveness of a super-market is judged in terms of the number of customers which can park their cars, move efficiently through the shop, through the check-out without waiting and return to their car and remove it to allow space for the next customer. This equation is greater than just providing the lowest cost building.

The time interest of a client may be judged in terms of the time to sell or time to redevelop. These may be shorter than the life of the building in terms of its structure and fabric.

10.5   THE TIME VALUE OF MONEY

There are two ways of bringing life-cycle costs to a comparable time base; present value, and annual equivalent.

Present value

Very simply, present value represents the amount needed to be invested today to pay for capital cost plus future running costs. The sum to be invested will be less than the sum of all the costs because some of the life-cycle costs will occur in the future and therefore the sum invested today will attract interest until the time when it is spent.

In other words, ignoring inflation, if the bank was offering 4% interest then how much would be required to be put in the bank today to buy a £6000 heating plant in 10 years time; obviously less than £6000 because of the effect of compound interest. Remember, **no inflation** therefore the heating plant stays the same price, the 4% interest rate can be viewed as the difference between the current bank base rate and the inflation rate.

The sum to be invested can be determined by reference to the present value of £1 table in any book of valuation and conversion tables:

Rate per cent = 4   Years = 10   Factor = 0.6755642

Therefore £6000 × 0.6755642 = £4053.39

therefore £4053.39 needs to be invested today to pay for a £6000 heating plant in 10 years time.

Annual equivalent

Annual equivalent is easiest thought of as the loss suffered by investing a sum of money in a building rather than a bank. The annual cost of investing in a building is the interest which would have been gained and spent per annum ie. not compounded. Therefore if £30 000 is spent on a building and interest is at 4% then £1200 p.a. is lost.

The problem with investing in a building rather than the bank is that at the end of the building's life all that is left is a crumbling ruin whereas if the money was invested in the bank the money would still be there. So if spending £30 000 on a building means losing the building at the end of its life, the loss is greater than £1200 p.a. It is £1200 p.a. plus the amount that would need to be invested each year to replace £30 000 at end of the building's life. This accumulating amount is termed a sinking fund.

In the above example the annual amount to be invested at 4% to recoup £30 000 after 60 years is 30 000 × 0.0042018 = £126.05

The factor of 0.0042018 again is found by reference to the tables. Total annual equivalent in this calculation is:

$$30\ 000 \times 4\% \quad = \quad 1200.00$$
$$+ \text{ a.s.f.} \qquad\quad = \quad \underline{\ \ 126.05}$$
$$\qquad\qquad\qquad\qquad 1326.05$$

The formulae

It is useful to consider how the formulae for discounting are derived and thereby consider the use of formulae direct instead of using the formulae translated into tables. This approach is particularly useful where a value management exercise may demand consideration of, say, a monthly interest rate or other time periods which are not commonly represented in the tables.

In this respect also it is worth considering the proofs of the various life-cycle cost calculations which are all derived from the compound interest principle.

Proof 1 – Compound interest

Compound Interest P = principal
i = interest expressed as a decimal
n = number of time periods
A = accumulated amount

It should be noted that i must be quoted in the same units as n, ie. if i = 0.1 per annum (10% per annum) then n must be quoted in years. Similarly if i = 0.025 per month (2.5% per month) then n must be quoted in months.

At beginning of first time period	$A = P$
At end of first time period	$A = P + Pi$
	$= P(1 + i)$
At end of second time period	$A = P(1 + i) + P(1 + i)i$
	$= P(1 + i)(1 + i)$
	$= P(1 + i)^2$
Similarly at end of the period n	$A = P(1 + i)^n$
Compound interest equation is	$A = P(1 + i)^n$

Proof 2 – Present value

In the above equation P = the initial capital which with the interest rate and the time periods are used to calculate the accumulated amount. Sometimes, as in the Present Value example earlier, the future accumulated value is known and what is desired is the value as at today's date of that accumulated value. This is termed **present value**.

P = Present value

if $A = P(1 + i)^n$

Re-arranging for P:

Then the Present Value equation is: $P = \dfrac{A}{(1 + i)^n}$

Proof 3 – Sinking fund

The sinking fund equation answers the question: what is the amount to be invested at the **end** of each time period to give a required sum at a required rate of interest? For example, what amount do I need to put into the bank at the end of each year following my decision to start a sinking fund, to accumulate enough money to buy a car in three years time?

P = principal
i = interest expressed as a decimal
n = number of time periods (years, months, etc)
A = accumulated amount
R = end of period payments

An instalment at end of the first year will earn zero interest as the money is deposited at the end of the first year. This can be proved as follows: From the compound interest equation:

$A = R(1 + i)^{n-1}$
$A = R(1 + i)^0$
$A = R$

The first instalment made will gain the most interest as it lies in the bank for the longest period of time. However, it will lie in the bank for one year less than the period between making the decision to commence a sinking fund and the date upon which the whole fund is withdrawn. Therefore, assuming a period of n years between making the decision and withdrawal, the first instalment at the end of n years, will earn interest of $n-1$ years and grow to:

$R(1 + i)^{n-1}$

In other words, the first instalment will attract the most interest because it has been invested for the longest period of time.

The second instalment at the end of n years will earn interest for $n-2$ years and grow to:

$R(1 + i)^{n-2}$

Therefore the total accumulated amount A is a geometric series

$A = R(1 + i)^{n-1} + R(1 + i)^{n-2} \ldots + R(1 + i)^2 + R(1 + i) + R$

$A = R(1 + (1 + i) + (1 + i)^2 + \ldots(1 + i)^{n-2} + (1 + i)^{n-1})$

Multiply both sides by $(1 + i)$

$A(1 + i) = R((1 + i) + (1 + i)^2 + (1 + i)^3 \ldots + (1 + i)^{n-1} + (1 + i)^n)$

$A + Ai = R((1 + i) + (1 + i)^2 + (1 + i)^3 \ldots + (1 + i)^{n-1} + (1 + i)^n)$

But:

$A = R(1 + (1 + i) + (1 + i)^2 + \ldots(1 + i)^{n-2} + (1 + i)^{n-1})$

Therefore:

$R(1 + \cancel{(1 + i)} + \cancel{(1 + i)^2} + \ldots \cancel{(1 + i)^{n-2}} + \cancel{(1 + i)^{n-1}}) + Ai =$

$R(\cancel{(1 + i)} + \cancel{(1 + i)^2} + \cancel{(1 + i)^3} \ldots + \cancel{(1 + i)^{n-1}} + (1 + i)^n)$

$R + Ai = R(1 + i)^n$

$Ai = R(1 + i)^n - R$

$Ai = R((1 + i)^n - 1)$

$$R = A\, \frac{i}{(1+i)^n - 1}$$

This could be expressed more elegantly as:

$A = R(1 + (1 + i) + (1 + i)^2 + \ldots\ldots(1 + i)^{n-2} + (1 + i)^{n-1})$

$$= R\sum_{s=0}^{n-1}(1 + i)^s$$

From the general result for the sum of a geometric series:

$$A = \frac{R(1-(1+i)^n)}{1-(1+i)}$$

$$= \frac{R(1-(1+i)^n)}{-i}$$

$$= \frac{R((1+i)^n - 1)}{i}$$

Re-arranging for R:

The sinking fund formula is: $R = A \dfrac{i}{(1+i)^n - 1}$

The sinking fund formula can be used to illustrate the calculation of the £126.05 given earlier, being the amount required per annum to accumulate £30 000 at 4% in 60 years.

$$R = 30\,000 \, \frac{0.04}{(1.04)^{60} - 1}$$

$$R = 30\,000 \, \frac{0.04}{10.47 - 1}$$

$$R = 30\,000 \, (0.0042239)$$

$$R = 126.716$$

The apparent difference between the results is caused by the fact that a calculator or computer works to a greater number of places of decimals than the tables and as such is more accurate.

Proof 4 – Mortgage

The loan repayment formula is derived by substituting the first equation (compound interest):

$$A = P(1 + i)^n$$

into the sinking fund formulae. Therefore:

$$R = P(1+i)^n \left(\frac{i}{(1+i)^n - 1} \right)$$

The mortgage formula is:

$$R = P\frac{(1+i)^n i}{(1+i)^n - 1}$$

This formula may be adjusted to calculate the monthly repayment of a mortgage at an annual rate of interest ie. the adjustment of the outstanding balance is carried out annually but the repayments are calculated monthly. This is done as follows:

$$R = P \, \frac{(1+i)^n i}{12((1+i)^n - 1)}$$

Example:

$$15000 \text{ at } 14\% \text{ over } 25 \text{ years}$$

$$R = 15\,000 \frac{(1.14)^{25} \times 0.14}{12((1.14)^{25} - 1)}$$

$$R = \frac{26.4619 \times 0.14 \times 15\,000}{12 \times 25.4619}$$

$$R = \frac{3.704668213 \times 15\,000}{305.5429897}$$

$$R = 181.87 \text{ per month}$$

Proof 5 – Years' purchase or present value of £1 per annum

The present value of £1 per annum represents a lump sum deposited in a bank earning interest, and from which a fixed amount may be withdrawn at the end of each year.

The formula is derived from an adaptation of the mortgage formula.

$$R = P \frac{(1+i)^n i}{(1+i)^n - 1}$$

Therefore:

$$P = R \frac{(1+i)^n - 1}{(1+i)^n i}$$

This formula can be further developed to that which commonly appears in other references by:

$$P = \frac{R}{i} \left(\frac{(1+i)^n}{(1+i)^n} - \frac{1}{(1+i)^n} \right)$$

$$P = \frac{R}{i} \left(1 - \frac{1}{(1+i)^n} \right)$$

$$P = \frac{R(1 - (1+i)^{-n})}{i}$$

Which gives the present value of an amount R per period of time (annum) for n periods (years) at i interest while allowing a sinking fund (ie. interest) to accumulate at the same rate of interest.

$$P = £1 \frac{(1-(1.06)^{-5})}{0.06} = 4.2124$$

It means that if I want to invest a sum of money today which will give me £1 per year for the next five years, how much do I need to put in the bank at 6% interest?

Test

End year 1	4.2124 × 1.06	=	4.46 – 1 = 3.46
End year 2	3.46 × 1.06	= 3.66 – 1	= 2.66
End year 3	2.66 × 1.06	= 2.81 – 1	= 1.81
End year 4	1.81 × 1.06	= 1.91 – 1	= 0.91
End year 5	0.91 × 1.06	= 1.00 – 1	= 0.00

This is the equation to use as a means of assessing the present value of an annual expenditure.

10.6   INFLATION

In all the formulae and examples given previously inflation has been ignored, which obviously in the real world cannot be the case. The effect of inflation is to decrease the value of money ie. the capital invested and the interest gained. Inflation can be taken account of in the following manner:

from the compound interest equation

$$A = P(1 + i)^n$$

but where t = bank base rate
f = inflation rate
i = the effective rate

then i is a function of both t and f.

The value of P in present money terms, ie. purchasing power, decreases geometrically with inflation rate f, that is:

$$A = \frac{P}{(1 + f)^n}$$

Similarly, the value of P in present money terms increases geometrically with bank interest rate t, that is:

$$A = P(1 + t)^n$$

Therefore, in the presence of both an interest rate t and an inflation rate f, A depends upon the ratio of $(1 + t)$ to $(1 + f)$:

$$A = P \frac{(1 + t)^n}{(1 + f)^n}$$

In order to determine the effective discount rate i which accounts for both t and f, the simple compound interest formula applies, that is:

$$A = P(1 + i)^n = P\frac{(1 + t)^n}{(1 + f)^n}$$

Therefore:

$$(1 + i)^n = \frac{(1 + t)^n}{(1 + f)^n}$$

$$(1 + i) = \frac{(1 + t)}{(1 + f)}$$

$$i = \frac{1 + t}{1 + f} - 1$$

For example, to calculate the net effect when inflation = 8% and bank borrowing = 12%

$$i = \frac{1.12}{1.08} - 1$$

$$i = 0.037 = 3.7\%$$

The net rate which we should use in discounting is 3.7%.

Example

If the above interest rates are assumed then what is present value of £30 000 in ten years.

$$P = \frac{30\,000}{(1 + 0.037)^{10}}$$

$$P = 30\,000 \times 0.6950$$

$$P = £20\,850$$

Test

If £20 850 is invested in the bank at 12% compound interest, the amount saved after 10 years:

$$A = P(1+i)^n$$

$$A = 20\,850(1.12)^{10}$$

$$A = £64\,757$$

In that same 10 year period, the item being saved for will increase in price by 8% compound.

$$A = 30\,000(1.08)^{10}$$
$$A = 30\,000 \times 2.1589$$
$$A = £64\,768$$

10.7   CAPITAL AND TIME

An economic evaluation is concerned with money and an evaluation over a period of time should take account of the time value of money. The value of money is affected in two ways.

Firstly, inflation. As has been demonstrated above, inflation is a rise in the general price level reflecting a decline in the purchasing power of money. The argument regarding inflation is long and complex. Some authors say that it can be ignored altogether based on the reasoning that income also goes up with inflation. Inflation can have an effect when the rates of inflation for differing items are not constant, eg. the price of gas may be rising at a rate in excess of the general inflation rate. Generally speaking, a rate based upon the difference between the bank base rate and the inflation rate should give a satisfactory rate for comparative calculations.

Secondly, real opportunity cost of capital. This is the real rate of return available on the best investment. Therefore if £1000 would earn £300 in one year if invested in a particular manner, then the opportunity cost of that £1000 is 30%. The real opportunity cost of capital can only be determined by reference to the client, and if significant may result in this rate being used as the discount rate instead of the rates illustrated above.

10.8   STANDARD FORMULAE FOR LIFE-CYCLE COSTING EXERCISES

Symbols

P = Principal or present value;
i = Interest rate as a decimal in time period units;
n = Time periods (usually years);
A = Accumulated amount;
R = Repayment (end of period payments);
t = Actual discount rate (usually bank base rate);
f = Rate of inflation.

Compound interest

$$A = P(1 + i)^n$$

Present value of £1

$$P = A/(1 + i)^n$$

Sinking fund

$$R = A * (i/((1 + i)^n - 1)$$

Loan repayment – Mortgage

$$R = P * ((1 + i)^n * i)/((1 + i)^n - 1)$$

Present value of £1 per annum

$$P = R * (1 - (1 + i)^{-n})/i$$

Inflation adjustment

$$i = ((1 + i)/(1 + f)) - 1$$

10.9   DATA FOR LIFE-CYCLE COSTING

Life-cycle costing can be carried out at a range of levels of quantification. At one extreme the analysis is carried out in the absence of quantitative data and is based upon the opinions of members of the value management team. Even at this stage useful exercises can be carried out. An understanding of the movements in interest rates and the importance of the clients intentions with relation to time can be determined by modelling on very sketchy data. At the other extreme, accurate calculations, usually with relation to parts of the design, can be completed using sound data.

Data comes in three forms:

- **Historic data**

  Historic data is recorded by estate managers, office managers, facilities managers and others whose job is concerned with the running of a building. These managers will record the costs of energy, cleaning and maintenance. If properly recorded this is a good source of life-cycle costing data. Often, however, this data is combined for accounting purposes and is difficult to obtain in its component parts. Building Maintenance Information Ltd., (formerly BMCIS) keep records of this type for a number of types of buildings.

  There is a feeling amongst some value engineers in the US that historic data is best converted into constants. There is the presumption that to know that it costs £10 000 per annum to clean a 2000 square metre office is less useful than knowing that in an office, a cleaner can vacuum 100 square metres of carpet per hour. The Washington practice of Smith, Hinchman and Grylls take this attitude, and publish life-cycle costing data in this form. [Life-cycle Cost Data by A.J.Dell'isola and S.J.Kirk (1983)].

- **The best available guess**
  Under this heading data is available from two sources. Firstly, the life and maintenance of components of construction can be obtained from manufacturers and suppliers. Information can usually be obtained in terms of ranges of life, for instance 'These fans work for years, they come with a two year guarantee but providing they are well maintained will run for 8–12 years no bother. We've some which are still going after 16 years'. From this comment it could be assumed that the fan is unlikely to fail in the first 2 years, is unlikely to last 16 years and has an average life of about 10 years. This is useful data. Data of this sort may also be obtained by reference to trade magazines.

- **Predictive calculations**
  The third source of data is from predictive calculations. An example of this the energy requirement for the heating of a building from which an annual heating bill may be determined.

  A formula for net seasonal (September to May heating season) energy requirement in kWh is:

$$\frac{\text{Design Heat Loss} \times 24 \times D \times K}{\text{Design Temp Differential} \times 1000}$$

D = Number of degree days (about 2100 for London and 2500 for Central Scotland)
K = Correction factors measured as the product of the working week (5 days = 0.8), the response time of the building (about 0.85 where there is a night shutdown), and the correction factors for occupancy and thermal capacity. (usually 1.0 for an 8 hour day under normal conditions). In this case K = 0.68.

Fabric design heat loss in W = Area of heat loss surfaces × U value in W/m² °C plus ventilation design heat loss.

$$\text{Ventilation design heat loss in W} = \frac{N \times \text{Vol} \times \text{Design Temp Diff} \times 1440}{3600}$$

where;

$$N = \text{number of air changes per hour}$$
$$\text{Vol} = \text{volume of the building in m}^3$$

Design Temperature
Differential = maximum difference between outside and inside temperatures catered for in the design.

The results of these calculations can only give a very approximate figure of use only for comparisons. The calculation above ignores entirely the choice of plant and plant efficiency but will allow some useful comparison of building geometry and choice of components with respect to energy.

10.10   VALUE MANAGEMENT AND LIFE-CYCLE COSTING

In undertaking value management studies it is important to know where the elements of life-cycle costs lie and their significance one with another. The following is an analysis of the prime life-cycle costs of a building.

10.11   DESIGN FACTORS IN LIFE-CYCLE COSTING

Maintenance

Various studies into maintenance have shown that maintenance costs constitute approximately 12% of the occupancy costs of offices, 7% for homes for the aged and 30% for laboratories. Maintenance is significant but not the highest priority.

BS 3811:1964 defines maintenance as 'a combination of any actions carried out to retain an item in, or restore it to, an acceptable condition'. An acceptable condition has to bear in mind the age of the component and it must be realized that eventually a component must be replaced. The costs reflected in the above figures are for maintenance only and do not include a sinking fund for future replacement. In a life-cycle costing calculation this sinking fund would be presumed where the product needed replacement during the time horizon.

In design terms a decision must be taken as to the choice of components in terms of their maintenance. If it is intended that components will be replaced during the life of the building then that must be designed to be physically possible at minimum cost. If the design presumes that components will be replaced there is a good argument for advising the client to set up the appropriate sinking fund and also to supply a maintenance manual for the building.

Energy

Energy costs contribute about 20%–35% of the operating costs of building the actual amount dependent on type, use and location.

Energy has seven life-cycle cost components:

1. The capital cost of the heating, ventilation/air conditioning plant and the capital cost of electrical plant, e.g. lights, lifts, etc. The extent to which these are required is a direct consequence of a design decision.
2. The cost of operating plant. If the design presumes the employment of a services engineer to run the system then it must be realized that the client will be responsible for that person's employment for the rest of the life of the building.

3. The cost of routine maintenance. If plant is to be regularly serviced then good access is required. There is an argument to the effect that, for instance, high quality air conditioning fans with sealed-for-life bearings can be built in to the structure. When the fan fails in 10 or 15 years time a part of the structure will have to be removed to gain access with some inconvenience to the occupants of the adjoining space. This however, may be a cost effective solution since it allows the client to rent the space which would have been allocated to a plant room.

4. Cost of periodic replacement. All plant has a finite life and must be replaced. The choice of plant should be made with reference to the client's time horizon. For example, if a client has a time horizon of 15 years, is it better to supply plant with a life of 8 years costing £100 000 or to supply plant with a life of 12 years costing £140 000? In this example either will be replaced once so the former is the most economic option.

5. Choice of fuel. The choice of fuel is always a gamble as to which will be cheaper in the long run. There are no rules here. Multi fuel boilers such as those which can burn both gas or oil may be economic, especially where a gas tariff has different cost levels based upon demand.

6. Energy consumption. Energy consumption will be related to the extent to which the building is serviced, the geometry of the building and the insulating or heat gain qualities of the external fabric. All of these aspects can be the subject of predictive calculations.

7. The clients energy management policies. Decisions by the client to have an artificially lit, air conditioned building give scope for the minimization of external envelope. Conversely, a decision to maximize natural lighting and rely only upon natural ventilation will result in a different design strategy.

Cleaning

Between 10 and 20% of the operation costs of buildings is spent in cleaning. Cleaning is an aspect which should be taken into account at the design stage. If a building is to be naturally ventilated there is an option to install reversible windows for cleaning. Otherwise a decision has to be taken to rely on ladders (can the external fabric support them?) or install a roof track and cradle. Modern plastic coated cladding needs to be cleaned perhaps every two years depending on location.

The extent to which internal finishes require regular cleaning is a factor which should be taken into account at the design stage. Hard surfaces which can be washed, dried and polished in one operation with a large machine will cost less than identical surfaces in small areas in which the three operations have to be undertaken by hand. Carpets are generally the least expensive surface to clean in confined areas but where vacuum cleaning is possible.

Security

The need for security in all types of buildings continues to increase and the extent to which a building can be secure is a direct consequence of the design. For example, if the client intends to staff each external entrance then the restriction of these will reduce future operating costs. Passive security can be obtained by having emergency exits in public areas rather than at the end of a staircase in a building serviced by lifts. Other points would include the omission of rooflights and windows at vulnerable points in the building.

Operating costs in security are staff and electronic devices. Costs can be restricted by minimizing the requirement for staff, increasing passive security in design and maximizing the facility for electronic devices. Action taken after the completion of the building is generally expensive and unsightly.

There may also be a trade-off between investing in security and lower insurance premiums. An insurance company may be able to assist here.

Commercial rates

Rates are payable by the occupiers of most buildings and account for the highest proportion of the operating cost of a building, 15%–45%. The principal determinant of rates is location. If the site is not fixed then rates should be taken into account. Rating is a particularly complex subject and the inclusion of a rating surveyor on the value management team should be considered.

Factors which would affect rates for a particular project are, for example, whether an office serving a factory is in a separate block or contained within the factory; whether a car park for an office is sufficiently separated such that it could be let separately and whether plant is internal or external.

Tax – capital allowances

A complex subject, but again the inclusion of a surveyor skilled with capital allowances may be a useful addition to value management teams dealing with certain commercial projects.

10.12 EXAMPLES IN LIFE-CYCLE COSTING

The following examples illustrate in various degrees of difficulty the application of life-cycle costing techniques. The examples have been chosen to demonstrate the importance of correct logic in the framing of a life-cycle cost problem.

1. *Example:* What is the maximum amount that it is economical to spend today to avoid a replacement costing £10 000 in 6 years time? Assume a discount rate of 7% and ignore inflation.

    Answer: Present value $= A/(1 + i)^n$

    $$= £10\ 000\ /\ (1.07)^6$$

    $$= £6663$$

In this simple example the result is that it is worth spending £6662 or less today to avoid the replacement. If the sum to be expended today is £6664 or more, then it is worth banking the money to be expended and buying the replacement in 6 years time.

2. *Example:* What is the maximum amount that it is economical to spend today to save £500 per annum over 10 years at 7% interest ignoring inflation.

    Answer: Present value $=$ Annual cost $\times$ P.V. of £1 p.a.

    $$= R * (1 - (1 + i)^{-n})/i$$

    $$= £500 * (1-(1.07)^{-10})/0.07$$

    $$= £500 * 0.49165\ /\ 0.07$$

    $$= £3511.79$$

This is an illustration of the classic insulation problem. It is worth spending £3510 or less on insulation to reduce fuel bills by £500 per annum. If the insulation costs more then from an economic viewpoint it is less expensive to burn fuel.

3. *Example:* What is the present value of a heat pump installation in a 130m² detached house given the following:

Initial purchase	= £1500
Annual maintenance	= £50 pa
Compressor replacement in 8th year	= £400
Annual electricity costs	= £800 pa
Time horizon	= 15 years
Residual value	= £250
Interest rate	= 7%

Answer.

Initial purchase	= £1500
Annual Maintenance	
£50 * $(1 - (1.07)^{-15})/0.07$	= £ 456
Compressor replacement in 8th year.	
£400/$(1.07)^8$	= £ 233
Annual electricity costs.	
800 * $(1 - (1.07)^{-15})/0.07$	= £7286
	£9475
Less Salvage £250/$(1.07)^{15}$	= £ 91
TOTAL	£9384

The above presumes that the heat pump can be funded out of existing capital. If this is not the case and borrowing is required then the following calculations should be undertaken.

Assume a mortgage for house improvement at 14%.

$$R = P * i * (1 + i)^n/ ((1 + i)^n - 1)$$

Assuming tax relief on the mortgage interest at 25% then an interest rate of 14% is effectively 10.5%.

$$R = 1500 * 0.105 * (1.105)^{15} / ((1.105)^{15} -1)$$
$$= 203 \text{ p.a.}$$

Present value of annual mortgage payment is:

$$P = 203 * (1 - (1.07)^{-15})/0.07$$
$$= 1849$$

The present value of the purchase on this basis is £1849 as opposed to the £1500 in the example.

The heat pump example is not a useful exercise unless the heat pump is an option amongst a number of options. In comparing technical alternatives note should be taken of the following decision tools.

10.13 LIFE-CYCLE COSTING DECISION TOOLS

Generally

Life-cycle costing is an exercise which is normally carried out in order to compare the total cost of two or more solutions to a design problem. In order to select the best method for carrying this out, the following rules should be observed.

Rule 1

In a situation where the project has to be undertaken the comparison is based on a total life-cycle cost calculation.

Rule 2

In the situation where the project **may** be undertaken the comparison is based upon the measures of economic performance.

Rule 1 applies therefore to situations of new build or to situations in which a replacement must be made through the failure of the existing component or system.

Rule 2 applies where the existing component or system is functioning satisfactorily but where investment in a retrofit may improve the current situation, eg. the installation of insulation. In order to judge between the relative merits of differing investment alternatives, a number of measures of economic performance have been developed.

In order to illustrate the application of the various methods, example number three above will be compared to an assumed existing oil fired central heating installation. It should be noted that the oil fired system is functioning satisfactorily.

Existing oil fired system

The existing oil fired central heating costs are as follows:

The annual fuel cost is £1100.

The annual maintenance cost is £100. (Assume this covers replacements).

Assume a life in excess of 15 years.

The present value of the cost of the oil fired system is:

$$P = 1200 * (1-(1.07)^{-15})/0.07$$
$$= 10\,929$$

Decision Tool 1 – Net savings

The first and most easily understood measure of economic performance is the net savings. The net saving is the difference between the amount invested and the amount saved. In the heat pump vs oil situation above the net savings by investing in a heat pump would be:

£10 929 – £9384 = £1545.

Decision Tool 2 – Savings to Investment Ratio (SIR)

$$\text{SIR} = \frac{\text{Reduction in (Energy Cost + Maintenance)}}{\text{Purchase} - \text{Salvage} + \text{Replacement.}}$$

In heat pump example.

$$\text{SIR} = \frac{10\,929 - 7742}{1500 - 91 + 233}$$

$$= \frac{3187}{1642}$$

$$= 1.94$$

Any SIR over 1.0 is worthwhile. This is a useful measure when comparing a series of options. Assume that a choice exists between, say, loft insulation, double glazing and cavity fill, but only one option can be afforded. The SIR will assist in the determination of the choice between the options, or alternatively will allow the options to be ranked on the basis of value for money.

Decision Tool 3 – Internal rate of return (IRR)

This gives the % rate of return on the investment. Unlike the preceding techniques the IRR does not require a discount rate, since it solves for a rate of interest. If this rate of return is acceptable to the investor then the project could go ahead.

The method is, by trial and error, to solve the equations for varying rates of interest. It should be noted that this method is prone to error since the solution assumes a straight line in an equation which is of a curve. See diagram 10.1 below. For this reason the differences between the rates of interest should be as small as possible and never greater than 5% apart.

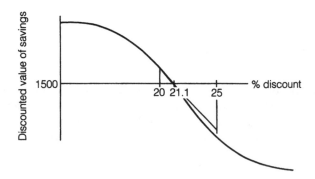

Figure 10.1   Varying results of present value at different interest rates.

The calculation is as follows:

		20%	25%
1.	PV of energy and maintenance savings £350 p.a.	1636	1350
2.	Investment	1500	1500
3.	PV of salvage (£250)	16	9
4.	PV of replacement (£400) in year 8	93	67
	Total 1 – (2 – 3 + 4)	59	–208

Therefore: 20% < IRR < 25%

by interpolation IRR $= 0.20 + 0.05 * \left(\dfrac{59}{59 + 208} \right)$

$= 21.1\%$

Decision Tool 4 – Simple Payback (SPB)

To give a quick appreciation of the viability of the scheme.

$$\text{SPB} = \frac{\text{Project First Costs}}{\text{Yearly Savings}}$$

$$= \frac{1500}{1200 - 850}$$

= 4 year 3 months

Decision Tool 5 – Discounted Payback (DPB)

This gives a more realistic appreciation of the problem by taking into account the effect of interest rates. It is a measure of the time taken before the accumulated savings offset the initial costs. It should not be used as an alternative to the other methods particularly in terms of comparison since the method does not permit a full appreciation of periodic maintenance or total life.

Method:– use SPB first to get an appreciation of break even point. Say from 4 years on.

Years into Investment	PV of Energy Savings	PV of Initial and Replacement Costs	PV of Salvage
1	–	1500	–
4	1186	0	0
5	1435	0	0
6	1668	0	0

DPB greater than 5 years, less than 6.

Decision Tool 6 – Sensitivity analysis

This is the area in which considerable work is still to be done. Sensitivity is the reaction to changes in the time horizon, the interest rates, the maintenance amounts etc., ie. variations in the variables. This to date is carried out by 'what if' questioning, ie. continually varying the variables and noting the change in the answer. This must be done where interest rates are likely to move.

Decision Tool 7 – Risk and uncertainty

Risk is defined as applying to a situation where the event is understood and the probability of the risk can be defined. An example would be the effect of the weather on construction where the effects of a day of continual rain is well understood and the probability of a day of continual rain within the construction period can be estimated. Uncertainty on the other hand is defined as the position when neither the event nor the probability can be estimated.

Risk is available in two forms, that which may be insured against and that which may not. Where a risk can be insured against, eg. a maintenance contract, then it is argued that the premium is a sensible basis upon which to base a life-cycle cost calculation. Where risk may not be insured against then risk management strategies should be adopted. Two basic methods, fully described in Flannagan *et al.* (1989), are outlined here.

The first method entitled ENPV (Expected Net Present Value) makes use of probability in order to arrive at the most likely figure to put into a life-cycle cost calculation. For example:

	Optimistic		Most likely		Pessimistic	
	Actual	Disc	Actual	Disc	Actual	Disc
Initial costs £	15 200	15 200	16 500	16 500	19 000	19 000
Annual costs (discounted 6%) Year 1 Year 2 Year 3	350 350 400	330 311 336	500 500 550	472 445 462	700 700 800	660 623 672
Net Present Value		16 177		17 879		20 955
Probability		0.25		0.50		0.25

It should be noted that the sum of the estimated probabilities must equal 1. In the example the ENPV is determined by the equation:

ENPV = 0.25 × £16 177 + 0.5 × £17 879 + 0.25 × £20 955

ENPV = £18 222.50

The second method, described under the heading of a risk management system, uses sensitivity analysis as a means for demonstrating the risk in terms of the variation in parameters to the calculation. The risk analysis is assessed by the construction of a spider diagram below. The flatter the slope on the diagram the more sensitive the LCC calculation is to variations in the parameter.

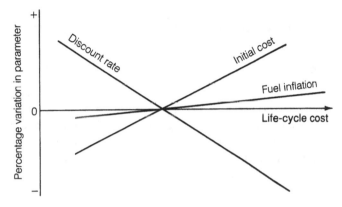

Figure 10.2   Spider diagram.

Decision Tool 8 – Discounted cash flow

The concept of discounted cash flow is useful within a life-cycle cost calculation in that it demonstrates that projects can be brought on stream at different points in the time horizon as cash becomes available, either out of budgets or from savings from previous schemes.

Within a discounted cash flow the money spent and/or available in each year of the time horizon is plotted. A spreadsheet is a particularly useful tool in this respect.

10.14   MORE COMPLEX LIFE-CYCLE COSTING EXAMPLES

Introduction

Life-cycle costing may be used in two situations; firstly in the investigation of options available for the improvement in the cost efficiency of an existing facility, or secondly in the evaluation of choices available at the stage of designing buildings or planning the replacement of worn components. The two situations require different approaches using common techniques.

Existing buildings

Exercises involving life-cycle costing can be prompted by either a breakdown in a system, in which there is a choice in the methods of rectification, or by a client or professional adviser suggesting that an initial investment in a project may save future running costs. An example of the latter would be the retrofit of double glazing, cavity wall insulation, etc., to save future energy costs.

Two examples illustrate the application of these points.

Example 1

In the first situation it is assumed that the company which has a maintenance agreement for the lifts in a government building have written stating that they are prepared to maintain the lift equipment for a further six years, after which time the lift motors will be beyond maintenance. This statement has prompted a study which has realized the following information.

The existing lift motors consume 2 million kWh of electricity per year. Present electricity costs are £0.06 per kWh. With extensive overhaul and modification costing £100 000 it is estimated that annual power consumption could be reduced by 10% and equipment life extended to 25 years. Without the overhaul the equipment is expected to last 6 years and overhaul will then not be possible.

New lift equipment is presently available at £400 000. It will cost £20 000 to remove and dispose of the old equipment. The new equipment is 25% more energy efficient than the current equipment without the overhaul and has an estimated life in excess of 25 years.

The maintenance contract for the lift equipment is £1000 per annum regardless of type of equipment.

The problem is now set and decisions need to be made on the appropriate life-cycle costing method. In this situation there is no choice, the motors have to be replaced. Therefore, the measures of economic performance are not applicable and the method of calculation will be based upon comparison of the total life-cycle cost of each option.

In order to carry out the calculation a number of assumptions need to be made. First, what is the time horizon? Since this is a government building it can be assumed that the client's interest in the building will be long.

However, study periods in excess of 20–25 years can be difficult in that the effect of discounting makes the variance between decisions almost meaningless. Since the overhaul will result in an estimated life of 25 years this time period will be used as the time horizon. Further assumptions need to be made with regard to interest and bank base rates. The following are assumed on the basis that in the next few years bank base rate will hover around 12%, the general rate of inflation will be around 5% and that the current trends in fuel prices will continue with fuel inflation at 3%.

The next stage is to set down the options to be investigated. In this case there are four:

1. Overhaul the lifts now;
2. Renew the lifts now;
3. Renew the lifts in six years time;
4. Renew at any point between now and six years.

Since 4 is a sub-set of 2 and 3 it will be ignored until the results of the first three options are available.

The first calculations required are to determine the discount rate for fuel and for items relating to the general inflation level.

Discount rates

$$\text{Discount rate for fuel} = \frac{1 + \text{Bank base rate}}{1 + \text{Fuel Inflation Rate}} - 1$$

$$= (1.12/1.03) - 1$$

$$= 0.0874$$

$$\text{General discount rate} = \frac{1 + \text{Bank base rate}}{1 + \text{General Inflation Rate}} - 1$$

$$= (1.12 / 1.05) - 1$$

$$= 0.06667$$

Option 1.   Overhaul now.

Overhaul cost   =   £100 000
Energy cost
2 million kWh – 10% = 1.8 million kWh
1.8 million × £0.06   = £108 000
Present value of £108 000 per annum for 25 years
$$= 108\,000 \times (1-(1.0874)^{-25})/0.0874$$

$$= 108\,000 \times 10.033$$
$$= 1\,083\,564 \qquad\qquad = \quad \underline{£1\,083\,564}$$

$$\text{TOTAL} \qquad \underline{\underline{£1\,183\,564}}$$

Option 2.   Replace now

New cost   =   £420 000
Energy cost
2 million kWh – 25% = 1.5 million kWh
1.5 million × £0.06   = £90 000
Present value of £90 000 per annum for 25 years

$$= 90\,000 \times (1 - (1.0874)^{-25})/0.0874$$
$$= 90\,000 \times 10.033$$
$$= 902\,970 \qquad\qquad = \quad \underline{£902\,970}$$

$$\text{TOTAL} \qquad \underline{\underline{£1\,322\,970}}$$

Option 3.   Replace in 6 years time.

Initial cost:
Present value of £420 000 in 6 years time

$$= 420\,000/(1.06667)^6$$
$$= 420\,000/1.4729245$$
$$= 285\,147 \qquad\qquad £285\,147$$

Energy cost for first 6 years:
Present value of £120 000 per annum for 6 years

$$= 120\,000 \times (1 - (1.0874)^{-6})/0.0874$$
$$= 120\,000 \times 4.521$$
$$= 542\,520 \qquad\qquad \underline{£542\,520}$$

$$\text{Carried forward} \qquad \underline{£827\,667}$$

Energy cost for years 6 to 25:
Present value of £90 000 per annum for 25 years

$$= 90\,000 \times (1 - (1.0874)^{-25})/0.0874$$
$$= 90\,000 \times 10.033$$
$$= 902\,970$$

Brought forward	£827 667

Less present value of £90 000 p.a. for 6 years

$$= 90\,000 \times (1 - (1.0874)^{-6})/0.0874$$
$$= 90\,000 \times 4.521$$
$$= 406\,890$$

Cost for years 6 to 25 = 902 970 − 406 890	£ 496 080
TOTAL	£1 323 747

Summary

Option 1.   £1 183 564
Option 2.   £1 322 970
Option 3.   £1 323 747

Therefore at the prices and rates given, option 1 is the best alternative. However, if the building has a life of in excess of 25 years, the client may wish to consider the residual value of the lifts at that time. Strictly speaking a sensitivity analysis should also be undertaken to demonstrate the effects of changing interest rates.

A further factor which should be considered is risk. Anything paid for today is obviously at minimum risk in terms of price fluctuations. The delay in the purchase of a new lift for six years results in some risk that prices may alter not in accordance with the assumed inflation rates.

Example 2

In this situation a retrofit option is considered. A client, an insurance company, is running a building which contains a computer suite. The computer suite is air conditioned. A consultant has suggested that the expelling of heat to the atmosphere is wasteful when the building has also a demand for domestic hot water. It is determined that using waste heat from its computer centre, currently 0.5 million kWh of recoverable heat can be used in heating water.

On the same theme the consultant has also suggested that installing self closing spray taps to all wash hand basins is estimated to reduce the domestic hot water consumption by 20%. Currently domestic hot water is heated by off-peak electricity at £0.04 per kWh and the company uses 1 million kWh per annum.

The cost of the first option is:

Install heat exchanger including all pipes and valves	£180 000

Additional maintenance per annum	£500
Re-tubing heat exchanger at 8 years	£15 000
Life of system	25 years

The cost of the second option is:

| Install spray taps | £12 000 |
| Maintenance per annum | £1000 |

Replacement of taps required at 6 year intervals.
The assumption for interest rates will be as follows:

Bank base rate	= 12%
Inflation rate fuel	= 3%
Inflation rate general	= 4%

In this case it has been decided to test for the sensitivity of the answer to changes in the above rates and therefore a short, simple computer program will be written to carry out the calculations.

This situation demonstrates how a variance in the parameters laid down for the study can change the result dramatically. The computer program enables the problem to be solved many times with different parameters. However, note that the computer will always solve the problem exactly and has taken for re-tubing in the 24th year even although this is unlikely to happen.

The program in BASIC is given below:

```
10 Lprint "Heat recovery exercise"
20 Lprint: Lprint
30 Input "Bank base rate ";b1
40 Input "Inflation rate fuel ";f1
50 Input "Inflation rate general ";g1
60 i=(1+b1)/(1+fl)−1
70 i1=(1+b1)/(1+g1)−1
80 Input "Yrs ";n
90 s=20000*((1+i)^n−1)/((1+i)^n*i)
100 Lprint "Fuel saving = ";s
110 Lprint : Lprint "Costs"
120 Lprint "Initial = £180000 "
130 m = 500*((1+i1)^n−1)/((1+i1)^n*i1)
140 Lprint "Maintenance = £";m
150 If n<8 then 210
160 x = INT(n/8)
170 For z = 1 to x
180 n1 = n1+8
190 rt=rt+15000/(1+i1)^n1
200 Next z
```

210 Lprint "Re-tubing = £ ";rt
220 Lprint "Total = £";180000+m+rt
230 End

For the heat recovery exercise the following results are given:

Bank base rate	Inflation fuel	Inflation general	Time yrs	Investment £	Saving £
12%	3%	4%	25	200 888	200 698
12%	5%	5%	25	203 485	240 241
12%	8%	7%	25	209 929	322 460
15%	8%	7%	25	201 400	244 376
12%	3%	4%	15	192 652	163 739

Having run the same problem a number of times it is now possible to give the client a number of scenarios.

1. The project is only viable if the time horizon is long, almost the life of the plant.
2. It is only viable at times when fuel inflation is above 5% with a bank base rate of 12%.

Further interpretations can be gained with more runs of the computer model. With each run taking about 30 seconds this is not a big time investment.

In the second part of this exercise the replacement of taps should be considered. The approach is the same, the following computer program will allow the investigation of economic performance under various interest rate and time conditions:

10 Lprint "Taps exercise"
20 Lprint:Lprint
30 Input "Bank base rate ";b1
40 Input "Inflation rate fuel ";f1
50 Input "Inflation rate general ";g1
60 i=(1+b1)/(1+f1)–1
70 i1=(1+b1)/(1+g1)–1
80 Input "Yrs ";n
90 s=8000*((1+i)^n–1)/((1+i)^n*i)
100 Lprint "Fuel saving = ";s
110 Lprint:Lprint "Costs"
120 Lprint "Initial = £12000"
130 m=1000*((1+i1)^n–1)/((1+i1)^n*i1)
140 Lprint "Maintenance = £";m
150 Lprint "Total = "; 12000+m
160 End

This exercise is one which demonstrates a worthwhile project under almost all circumstances. The repeated computer runs below give a positive position almost irrespective of the interest rates.

Bank base rate	Inflation fuel	Inflation general	Time yrs	Investment £	Saving £
12%	3%	4%	6	16 666	36 170
12%	5%	5%	6	16 816	38 528
12%	3%	4%	3	14 592	20 345
12%	0%	2%	3	14 496	19 215

A project of this kind, where even with no fuel inflation a saving results within three years, can be recommended to a client with very few reservations.

New Buildings – Example 3

In general, the life-cycle costing calculations for a new building will be carried out on the basis of comparing the total life-cycle costs of a number of design options. The design options will relate to components or elements and rarely be for a complete building.

Consider this example. In the design of an office block the designer requires cost advice on the choice of a type of partition. The partition system chosen will be required to be taken down and re-erected at intervals to reflect the client's changing requirements for office accommodation. The following represents the options to be considered.

1. Blockwork plastered and painted and built off the screed.

   Cost to erect   £25.00 per m^2

   Cost to take down   £12.00 per m^2

   Maintenance: paint every   £ 4.00 per m^2
   5 years

2. Stud partition with plasterboard dry lining and painted both sides.

   Cost to erect   £27.00 per m^2

   Cost to take down   £2.00 per m^2

   Average replacement of
   plasterboard due to damage   £2.00 per m^2 per year

   Maintenance: paint every
   5 years   £4.00 per m^2

3. Steel demountable partition system.

   Cost to erect   £90.00 per m^2

   Cost to take down & re-erect   £5.00 per m^2

   Maintenance: cleaning   £4.00 per m^2 per year

In this situation three things are important:

1. For what period has the client an interest in the building? Assume 15 years.
2. How often are office layouts to change? Assume 1, 2, 5 and 7 year intervals.
3. Discount rate? Assume 7% but test for variance.

Discount rate 0.07 Time horizon 15 years

Interval years	1	2	5	7
Blockwork				
Erect	25.00	25.00	25.00	25.00
T.d. & re-erect				
(P.V. £37 for)				
Each year	323.58			
Year 2		32.32		
Year 4		28.23		
Year 5			26.38	
Year 6		24.65		
Year 7				23.04
Year 8		21.53		
Year 10		18.81	18.81	
Year 12		16.43		
Year 14		14.35		14.35
Maintenance at 5 years				2.85
Maintenance at 12 years				1.78
Total	348.58	181.32	70.19	67.02

Interval years	1	2	5	7
Stud partition				
Erect	27.00	27.00	27.00	27.00
T.d. & re-erect				
(P.V. £29 for)				
Each year	236.13			
Year 2		23.58		
Year 4		20.60		
Year 5			19.25	
Year 6		17.99		
Year 7				16.81
Year 8		15.71		
Year 10		13.73	13.73	
Year 12		11.99		
Year 14		10.47		10.47
Replace damaged	17.49	17.49	17.49	17.49
Maintenance at 5 years				2.85
Maintenance at 12 years				1.78
Total	280.62	158.56	77.47	76.40

Interval years	1	2	5	7
Steel partition				
Erect	90.00	90.00	90.00	90.00
T.d. & re-erect				
(P.V. £5 for)				
Each year	43.73			
Year 2		4.37		
Year 4		3.81		
Year 5			3.56	
Year 6		3.33		
Year 7				3.11
Year 8		2.91		
Year 10		2.54	2.54	
Year 12		2.22		
Year 14		1.94		1.94
Maintenance cleaning	34.98	34.98	34.98	34.98
Total	168.71	146.10	131.08	130.03

Given the above the best choice is steel if the client is likely to re-arrange at intervals of two years or less. Exceeding this period blockwork is less expensive although the difference between this and stud is very little.

It should be noted that the example is simplistic and does not take account of disruption to the floor or ceiling finish, nor to services. Nevertheless it does provide a base point from which to commence discussion and these factors can easily be brought in if it is thought worthwhile.

Existing buildings – Example 4

A manufacturer wishes to reduce the running costs of his factory which are per annum:

Fuel (oil)	£18 000
Heater maintenance	£1000
Roof maintenance	£1500

He has also determined that fuel and maintenance savings could be achieved by undertaking one or more of the projects under:

	Project Cost	Fuel Saving
1. Install cavity wall insulation	£ 6000	£2500
2. Cover existing roof with insulated aluminium sheeting (no further roof maintenance will be required)	£30 000	£5500
3. Install insulation below existing roof	£14 000	£4500
4. Replace oil heaters with gas (heater maintenance will be reduced to £500 p.a.)	£25 000	£4500

Bank base rate   = 12%
Inflation rate fuel   = 4%
Inflation rate general = 3%

The factory owner also wishes to consider the implications of an initial spending limit of £30 000.

The use of the measures of economic performance are considered in the following pages. The use of a spreadsheet allows the data to be easily sorted into any rank order. The results demonstrate that the largest net saving is brought about by the largest investment but that the second largest net saving, only £900 less, is realized through an investment of £25 000 less. The best value for money is offered by the cavity wall insulation.

If it is assumed that the factory owner would like to take advantage of project 5, how can this be achieved within a spending limit of £30 000? The answer lies in a study of cashflow as illustrated below. However a discounted cashflow, carrying out the installation of cavity wall insulation at the end of year 6, demonstrates that the net saving and cashflow are inferior to a combination of cavity wall insulation and insulation below the roof.

Bank base rate	0.12
Fuel inflation	0.04
General inflation	0.03
Fuel discount	0.077
General discount	0.087
Time horizon	15 years

Projects	Invest	Saving	Net saving	S.I.R.	Simple Payback years
1. Cavity wall insulation	£6000.00	£21 806.67	£15 806.67	3.63	2.40
2. Cover roof	£30 000.00	£60 255.10	£30 255.10	2.01	4.29
3. Below roof insulation	£14 000.00	£39 252.00	£25 252.00	2.80	3.11
4. Oil to gas	£25 000.00	£43 345.48	£18 345.48	1.73	5.00
5. Cav wall & cover roof	£36 000.00	£82 061.77	£46 061.77	2.28	3.79
6. Cav wall & roof ins	£20 000.00	£61 058.67	£41 058.67	3.05	2.86
7. Cav wall & oil to gas	£31 000.00	£59 700.48	£28 700.48	1.93	4.51
8. Cover roof & oil to gas	£55 000.00	£91 606.91	£36 606.91	1.67	5.18
9. Below roof & oil to gas	£39 000.00	£72 784.48	£33 784.48	1.87	4.66
10. Cav wall & cover roof & oil to gas	£61 000.00	£107 961.91	£46 961.91	1.77	4.88
11. Cav wall & below roof & oil to gas	£45 000.00	£89 139.48	£44 139.48	1.98	4.39

Order by Net Savings

Projects	Invest	Saving	Net saving	S.I.R.	Simple Payback years
10. CW & CR & O–G	£61 000.00	£107 961.91	£46 961.91	1.77	4.88
5. CW & CR	£36 000.00	£82 061.77	£46 061.77	2.28	3.79
11. CW & O–G & Rf	£45 000.00	£89 139.48	£44 139.48	1.98	4.39
6. CW & Rf ins	£20 000.00	£61 058.67	£41 058.67	3.05	2.86
8. CR & O–G	£55 000.00	£91 606.91	£36 606.91	1.67	5.18
9. O–G & Rf ins	£39 000.00	£72 784.48	£33 784.48	1.87	4.66
2. Cover roof	£30 000.00	£60 255.10	£30 255.10	2.01	4.29
7. CW & O–G	£31 000.00	£59 700.48	£28 700.48	1.93	4.51
3. Below roof	£14 000.00	£39 252.00	£25 252.00	2.80	3.11
4. Oil to gas	£25 000.00	£43 345.48	£18 345.48	1.73	5.00
1. Cavity wall	£6000.00	£21 806.67	£15 806.67	3.63	2.40

	Projects	Invest	Saving	Net Saving	S.I.R.	Simple Payback years
		Order by S.I.R				
1.	Cavity wall	£6000.00	£21 806.67	£15 806.67	3.63	2.40
6.	CW & Rf ins	£20 000.00	£61 058.67	£41 058.67	3.05	2.86
3.	Below roof	£14 000.00	£39 252.00	£25 252.00	2.80	3.11
5.	CW & CR	£36 000.00	£82 061.77	£46 061.77	2.28	3.79
2.	Cover roof	£30 000.00	£60 255.10	£30 255.10	2.01	4.29
11.	CW & O–G & Rf	£45 000.00	£89 139.48	£44 139.48	1.98	4.39
7.	CW & O–G	£31 000.00	£59 700.48	£28 700.48	1.93	4.51
9.	O–G & Rf ins	£39 000.00	£72 784.48	£33 784.48	1.87	4.66
10.	CW & CR & O–G	£61 000.00	£107 961.91	£46 961.91	1.77	4.88
4.	Oil to gas	£25 000.00	£43 345.48	£18 345.48	1.73	5.00
8.	CR & O–G	£55 000.00	£91 606.91	£36 606.91	1.67	5.18
	Order by Pay back					
1.	Cavity wall	£6000.00	£21 806.67	£15 806.67	3.63	2.40
6.	CW & Rf ins	£20 000.00	£61 058.67	£41 058.67	3.05	2.86
3.	Below roof	£14 000.00	£39 252.00	£25 252.00	2.80	3.11
5.	CW & CR	£36 000.00	£82 061.77	£46 061.77	2.28	3.79
2.	Cover roof	£30 000.00	£60 255.10	£30 255.10	2.01	4.29
11.	CW & O–G & Rf	£45 000.00	£89 139.48	£44 139.48	1.98	4.39
7.	CW & O–G	£31 000.00	£59 700.48	£28 700.48	1.93	4.51
9.	O–G & Rf ins	£39 000.00	£72 784.48	£33 784.48	1.87	4.66
10.	CW & CR & O–G	£61 000.00	£107 961.91	£46 961.91	1.77	4.88
4.	Oil to gas	£25 000.00	£43 345.48	£18 345.48	1.73	5.00
8.	CR & O–G	£55 000.00	£91 606.91	£36 606.91	1.67	5.18

CAVITY WALL INSULATION & COVER ROOF

Year	Cover Roof	Cavity Wall Ins	Total Cashflow
1	−23 513.39		−23 513.39
2	−17 502.43		−17 502.43
3	−11 932.15		−11 932.15
4	−6770.17		−6770.17
5	−1986.47		−1986.47
6	2446.72		2446.72
7	6555.15	−3337.99	3217.16
8	10 362.67	−1956.14	8406.53
9	13 891.38	−672.99	13 218.39
10	17 161.73	518.51	17 680.24
11	20 192.69	1624.90	21 817.59
12	23 001.83	2652.26	25 654.09
13	25 605.41	3606.24	29 211.65
14	28 018.51	4492.08	32 510.59
15	30 255.10	5314.64	35 569.75

Year	Cashflow
1	−13 500.00
2	−7464.29
3	−1859.69
4	3344.57
5	8177.10
6	12 664.45
7	16 831.28
8	20 700.47
9	24 293.29
10	27 629.49
11	30 727.38
12	33 604.00
13	36 275.14
14	38 755.49
15	41 058.67

CHAPTER 11 ⎯⎯⎯⎯⎯⎯⎯⎯

GROUP DYNAMICS AND TEAM SKILLS

11.1   INTRODUCTION

This chapter outlines ideas, tools and techniques that are applicable for enhancing or developing team skills of use in a value management situation. The main themes to be covered are:

- Groups and teams;
- Group dynamics;
- Group problem solving skills;
- Negotiation;
- Chairmanship skills.

An understanding of team dynamics and the ability to be able to manage these processes is essential in assessing and analysing the client's situation.

11.2   GROUPS AND TEAMS

What is a group

Schein (1980) defines a group in psychological terms as any number of people who interact with one another; are psychologically aware of one another; and perceive themselves to be a group

This definition suggests there are limits to group size. For optimum activity, group size should be between six and ten people (Hunt 1986). Once size is extended to beyond ten people there is a danger that the group will fragment and small cliques will form. There are different types of groups.

Types of groups

Different types of groups can exist (Schein 1980). Major types are formal groups and informal groups.

Formal groups

Formal groups are those that are specifically created by managers to perform organizational tasks – work groups. There are two types of formal groups;

(a)   Permanent formal groups that exist with the knowledge that they are unlikely to be disbanded. Standing committees are one example. In this instance, however, individual group membership may change over time but the committee continues in existence.

(b)   Temporary formal groups. Of interest here are project teams and task forces. A value management team is an example of a temporary formal group.

Temporary formal groups can exist for long periods of time. The difference between the temporary and permanent groups is, however, not one of longevity of existence through time. It is the fact that members of a temporary formal group know amongst themselves that at some time in the future they will cease to exist as a group. In other words they have a life-cycle from formation through to cessation.

Informal groups

In a work related context informal groups will often emerge out of formal group activities but this may not always be the case. Thus an informal group is one that meets more for social reasons as opposed to organiza-tional task related reasons. Also, informal groups may emerge as a func-tion of the probability of interacting, for example, as a consequence of daily activities. In this instance it is often related to physical proximity, which increases the likelihood of social interaction. Whilst not generally a consequence of the need for social interaction in groups, informal groups can subvert on-going formal activities if they form an identity of their own which is counter to that of management's view of the world. In large organizations, informal groups can be broken down into the following types (Schein 1980):

1. Horizontal cliques, where group members are of near equal organizational status and may often work in close proximity.
2. Vertical cliques, where group members may be from different organizational status levels within the same department or team and know
   each other either from non-work-related activities or because they
   need to achieve particular goals relevant to common interests.
3. Mixed or random cliques, whose members are from different organizational status levels, departments and locations. This type of group
   may emerge to serve common interests or fulfil objectives that the
   formal organization is unable to meet.

Project teams as work groups

In the construction industry many activities are carried out in work groups
that have a finite and known life span and with a specific task or tasks to
perform – temporary work groups – usually given the connotation 'project
teams'. As Male (1991) has suggested, project teams in the construction
industry can either comprise people from within the same organization –
intra-organizational project teams – or from different organizations – inter-
organizational project teams; a design team being one example, value
management teams may comprise both. Project teams in the construction
industry have also been termed 'temporary multiple organizations' (Churns
and Bryant 1984) or project coalitions (Winch 1987). The former label
denoting the fact that project teams comprise representatives from different
organizations brought together to achieve a particular task, whilst the latter
label denotes the fact that power structures will also exist within project
teams between different organizational representatives.

The next section analyses the social processes that operate within
groups – group dynamics.

11.3   GROUP DYNAMICS

Group dynamics or processes within groups can be affected by seven
factors (Hellriegel, Slocum and Woodman 1986):

1. Group size, highlighted and discussed above.
2. Membership roles and composition, concerned with problem-solving
   skills and task orientated or socio-emotional behaviour patterns present within the group.
3. Group norms concerned with emergent implicit or explicit codes of
   behaviour shared by contributors.
4. Group goals. That is, the task to be performed or objectives attained.

5. Group coherence, which depends on the frequency of group meetings, the strength of the desire to be in the group and the importance of attaining objectives through task performance.
6. Leadership, both formal and informal.
7. External environment. This can be considered to be those factors outside of the group that impact its operations but over which it may have limited control.

These seven factors interact to provide an 'atmosphere' or 'climate' which is characteristic to each particular group. Hunt (1986) suggests that analysing group behaviour under the three generic headings of Group Norms, Roles, and Structure, provides a highly generalized framework allowing managers to understand group operations.

Group norms

Group norms are codes or rules of behaviour. These emerge from group interactions and the shared expectations of contributors and will either overtly or covertly control member behaviour. Group norms will take some time to emerge and therefore the longer the group is operating together the more likely that strong norms will develop. Hunt (1986) contends that the degree of conformity to group norms will depend on:

● A person's desire for the group to accept his or her membership – acceptance;
● A person's desire to avoid displeasure, punishment or isolation from the group – pleasure;
● A person's belief that group norms are a reflection of personal views – congruence;
● A person's ability to handle the doubt that they may not be able to stand alone – isolation;
● A person's belief in group goals – agreement.

The development of group norms also operate in the context of the structuring within a group.

Group roles

Each person has a set of expectations about how they as an individual will behave in social situations. Individuals also hold expectations, beliefs and assumptions about how other people will behave in social situations. These two sets of impressions combine to give an individual a situational view of the world, termed a role (Hunt 1972). Hunt (1986) indicates that studies of groups have found repeatedly two recurrent behaviour patterns:

1. Task-centred behaviour concerned with structuring, organizing, goal setting and often dominating group processes;
2. Socio-emotional or maintenance behaviour which is concerned with supportive, consensus seeking, conflict resolving, integrative processes.

Hunt suggests that contributors to the group will, over time, sort out these two types of behaviour amongst themselves in order to enable the task to be undertaken. Observations by Male in numerous simulated task-centred multi-disciplinary team situations also suggests that personal problem-solving styles, professional background and personality interact in a complex way in problem solving situations. These observations suggest that where there is a 'fit' between group dynamics and these variables the group is more concerned with task performance. However, if there is a degree of imbalance between these variables the group will spend more time on managing itself rather than concentrating on the task. If there is a severe imbalance, usually in the area of preferred problem-solving style and personality rather than professional discipline, the group can 'lock' into managing group processes almost at the total exclusion of task performance.

Group structure

In formal work groups team structuring is affected by hierarchy, authority and managerial style (Hunt 1986). There may be a formal leader appointed to manage the group but sometimes an informal leader(s) may also emerge in addition to the formal leader. Occasionally, depending on the balance between the perceived expertise of the formal leader and those who perform task-centred and maintenance behaviours, the formal and informal leaders may clash. Group norms will have a very strong guiding influence on who is recognized ultimately as the leader of the group.

The group development process

In bringing together the views of Adair (1987), Helriegel, Slocum and Woodman (1986) and Hunt (1986) a useful model for analysing the development of a group is outlined in Table 11.1.

As the group development process proceeds, and, depending on the balance of expertise, power distribution, problem solving styles and leadership behaviour coupled with the degree of group isolation from outside influences, a phenomenon called 'groupthink' can develop.

Table 11.1 A group development model

Phase	Comments
Forming	Getting to know each other.
	Task orientated behaviour is attempting to grapple with the requirements of task performance. Maintenance behaviours are concerned with socialization issues, overcoming anxieties and working out power relationships.
Storming	Power struggles emerge.
	Competition and conflicts between contributors emerge and the formal leader may be challenged. The management of conflict rather than its suppression becomes essential. Polarization of ideas can occur as can psychological withdrawal from the group. The feasibility of the task is questioned.
Norming	Rules of behaviour emerge implicitly or explicitly.
	Task orientated behaviour ensues in the form of the free-sharing of information, acceptance that opinions will differ and a search for compromise on tasks and objectives. Maintenance behaviours are directed towards establishing cohesion and support. Conflict is reduced and power is distributed. Ground rules are established and a sense of common responsibility for task performance as identified by the group emerges.
Performing	Task performance.
	Contributors' roles are clearly identified, understood and accepted. A frank exchange of facts, opinions and preferences occurs. Trust is established and problem solving is free-flowing, decision making occurs and the group experiences a high degree of cohesion. The group has worked out a structure to achieve the task or objective.
Adjourning	Task orientated behaviour is terminated.
	Disengagement from maintenance behaviour.

Groupthink

This is a term coined by Janis (1972) and discussed in more detail by Janis and Mann (1977) to explain what happens in highly cohesive, conforming groups. The characteristics of groupthink are:

- The illusion of invulnerability leading to over optimism and extreme risk taking;
- Group rationalization without re-validating assumptions prior to taking action;
- An absolute belief in the moral integrity of the group;
- Stereotyping of outsiders;
- Uncompromising pressure to conform to group norms;
- Uncritical thinking and self-censorship;
- The illusion of unanimity.

Hunt (1986), based on his own observations, adds an additional three characteristics:

- A belief that all contributors have expressed their views and that the outcome results from a consensus of divergent views;
- A concern for any answer regardless of its merits;
- A failure to identify expertise among contributors.

Summary

This section has reviewed group dynamics and identified the fact that understanding the emergence of group norms, roles and structure is important in being able to handle and manage group interactive processes, such as problem solving.

11.4   GROUP PROBLEM SOLVING SKILLS

The primary purpose of introducing this section in a chapter on groups, group processes and team dynamics is that value management procedures are undertaken using the collective wisdom and experience of a team. It is the value management team co-ordinator's function to unlock and subsequently guide the team's problem solving efforts towards the project. This section introduces and discusses a number of different ideas on problem solving and then subsequently proceeds to highlight a number of techniques that can be used to assist in this process.

This section avoids undertaking a treatise on creativity. It is viewed here as a mental process that is concerned with sifting, manipulating and reconfiguring information, is often subliminal but which can be stimulated consciously by tools, techniques and procedures. Creativity is not only an aid in but is essential to problem solving. The basic premise is that everybody is creative to varying degrees and that creativity has to be enhanced and channelled towards achieving the solution to a problem. In the context of value management this would mean analysing and improving the client's project.

Blocks to effective problem solving

Problem solving can be interrupted or fail to get started for a number of reasons. These can best be summarized as (Adams 1987):

1. Perceptual blocks, preventing the analyst from seeing the problem clearly or seeing the information needed to solve the problem. Within this mechanism are distinct categories of obstacles:
   (a) Stereotyping, where an individual has preconceived notions on a subject, attaches labels too readily or fills in information where none exists. This is a common problem with groupthink.
   (b) Difficulties in problem isolation due to inadequate contextual clues or misleading information.
   (c) Defining the problem area too tightly or imposing artificial limits or constraints.
   (d) An inability to adopt multiple perspectives.
   (e) Saturation by information from all sensory modes or, the converse.
   (f) An inability to use all sensory modes.

2. Emotional blocks, which are related to the basic psychological functioning of a human being in terms of needs and desires. Common types of emotional blocks are:
   (a) Fear of taking risks, the most general and common difficulty.
   (b) An inability to handle ambiguity, uncertainty and chaos.
   (c) Preferring to judge rather than generate ideas, primarily because it is safer to undertake the former than latter.
   (d) An inability to relax or suspend judgement and allow ideas to incubate, 'simmer' and ferment – one of the prerequisites for good problem solving.
   (e) A lack of motivation or too highly motivated to solve a problem.
   (f) An inability to discern unthreateningly the use of reality and fantasy, hence restricting the use of one or the other in problem solving.

3. Cultural and environmental blocks. The former is derived from exposure to societal patterning effects whilst the latter stems from the immediate social and physical environment.
   Cultural blocks include:
   (a) Taboos.
   (b) Denigration of: playfulness, the use of humour, fantasy, and reflective thinking in problem solving.
   (c) That the only good style to adopt for problem solving is rational thinking, as exemplified in the use of logic, numeracy, reason and

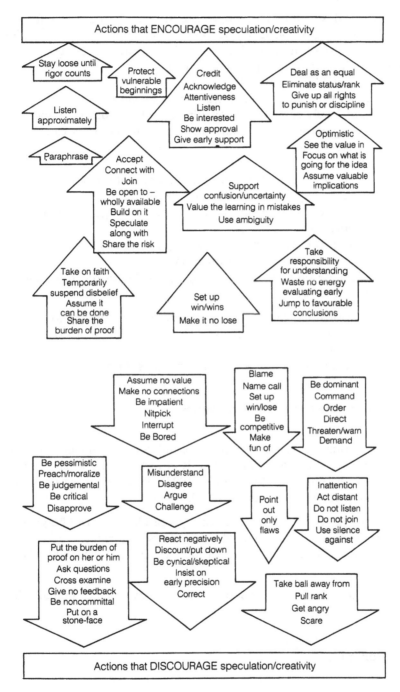

Figure 11.1 Speculation/Creativity. Source: Adams (1987), reprinted with permission.

practicality, – the 'hard' approach. An inappropriate style for problem solving is considered to be the 'soft' approach, as exemplified
by the use of intuition, feelings and qualitative judgement.
(d) The status quo is preferable to change.
Environmental blocks include:
(a) A lack of co-operation and trust among colleagues.
(b) Superior-subordinate relationships based on autocracy and a lack
of recognition of the contribution that subordinates can make to
problem solving.
(c) Physical distractions.
(d) Lack of management support for implementation of ideas.
4. Intellectual or expressive blocks. These stem from:
(a) Use of the incorrect problem-solving language – verbal, visual,
mathematical.
(b) Inappropriate use of problem solving strategies.
(c) Insufficient or incorrect information.
(d) Inadequate problem solving language skills for recording information and ideas.

One final problem solving hurdle is the particular learning style that an
individual has adopted through experience to tackle a problem, Honey
and Mumford (1982).
   The next section outlines tools and techniques for problem solving.

Problem solving tools and techniques

This section draws together a number of different techniques that can be
used in problem solving. It is a synthesis of material and ideas developed
by Adams (1987) and de Bono (1970; 1982). Two key aspects associated
with problem solving are (Adams 1987):

1. Fluency, that is, the number of concepts or ideas that can be generated
in any given amount of time. Fluency relates to quantity.
2. Flexibility, that is, the diversity of ideas generated. Flexibility relates to
variety.

Flexibility and fluency in generating ideas may not be mutually interrelated.

Brainstorming

Brainstorming is one of the most common techniques used for problem
solving and lateral thinking (Adams 1987, De Bono 1970). It works best

on simple problems or those that are well defined. The basic principles of brainstorming are:

- The quantity of ideas is important regardless of how wild they are;
- Team participants should be encouraged to build on or modify ideas;
- No evaluation of ideas should be permitted.

Adams indicates that the initial progress in a brainstorming session is rapid as contributors are able to provide common solutions. The problem for managing a brainstorming session is just beginning, however, since once the initial common pooling of ideas is used up the process becomes more difficult. It is precisely at this stage, when the group may flag, that an new impetus is required for novel solutions to emerge. In a value management context, this places considerable emphasis on the facilitator to ensure that group momentum continues and it becomes a test of that person's leadership skills to ensure group motivation remains high.

Synectics

Synectics appears particularly applicable to value management exercises since the presence of a 'client' is important for the process to operate. Synectics is more complex and sophisticated than brainstorming, it allows criticism and requires a higher level of technical expertise. The basic principles are set out in Fgure 11.1 (Adams 1987). Other factors are:
1. The client states the problem, is totally involved in the group processes and problem-solving exercise and selects the ideas from those presented during the session.
2. The team facilitator does not participate directly in solving the problem but is there to record ideas and maintain momentum.
3. The 'excursion' principle, used by the facilitator, involves:
   (a) The client stating a goal for which possible solutions are required. This provides the team with direction.
   (b) A key word reflecting an action or concept from the goal is selected by the facilitator.
   (c) The facilitator chooses a situation remote from the problem and asks the team to think of examples of that key word in relation to the situation or world totally remote from the context of the problem. This is asking the team to use fantasy and relax being rooted in reality. It also takes the team's focus away from the real problem.
   (d) The facilitator focuses the team's attention solely on one of the listed examples and asks them to explore the images and associations it conjures up. These associations and images are also recorded for latter use.

(e) Some or all of these associations are used to generate a small number of totally absurd ideas that are related back to the real problem. This step is, in effect, pulling the team back from its world of imagination and onto the problem in hand but attempting to use some of the fantasy images that have been conjured up to reframe the problem. It is also trying to overcome taboos, etc.

(f) Other ideas – Adam's calls them second generation ideas, are developed from one of the absurd ideas – extracting the key principles and rooting these back into the real problem situation without losing any of the underlying essence inherent in the absurd idea.

(g) The facilitator then asks the client to pick a second generation idea which is subsequently related back to the original problem. The client's responses are recorded.

(h) The process is repeated.

4. The focus is on producing fewer ideas than in brainstorming but relating these to solving the client's problem.

5. Criticisms are couched in the language of reservations and must be preceded by two positive statements directed at solving the client's problem.

Clichés and labels

De Bono (1970) argues that the patterning processes of the mind cause difficulties in problem solving. Clichés and labelling are areas that are particularly prone to narrowing an individual's attention down. He suggests that there are three approaches to challenging clichés and labels:

- Challenge the label by questioning why it is being used. Is the label essential or only convenient and does it have to be used only because others use it? This questioning process is not looking for a definition of the label or disagreeing with its use.
- Trying to do without labels and allowing the underlying 'essence' behind the label to emerge from the contextual information.
- Establishing new labels which reduces the distortion effect of the old labels and helps to prevent channelling new information into old labels.

The PMI technique

De Bono (1982) has suggested the use of PMI as an attention directing tool. PMI stands for:

P   Plus or good points
M   Minus or bad points
I   Interesting points, that may wish to be explored.

By using this simple technique of directing attention at each of these in turn, perhaps over a range of ideas, for a period of two to three minutes each the problem solver is adopting a scanning and objectivity mode of thinking.

The CAF technique

This is another attention directing tool suggested by de Bono (1982). The letters stand for Consider All Factors in a problem situation with no attempt at evaluation. This technique is particularly useful when considering undertaking a systems-thinking perspective on a project and exploring issues of what is to be taken account of in an 'environment'.

The TEC problem solving framework

This is a useful framework for focusing thought processes as a deliberate task (de Bono 1982). The letters stand for:
T   Target and Task of thinking. The target is the precise focus of attention, for example, a window and the task is the thinking task to be performed – identifying its functions.
E   Expand and Explore. This aspect of the framework is concerned with opening up the problem under investigation, for example, listing all the functions that a window performs and perhaps using visualization, as another problem solving technique, to enrich functional definition.
C   Contract and Conclude. This involves drawing things to a conclusion and arriving at a set of indentifiable outcomes.
    De Bono suggests that time periods should be allocated to each portion of the TEC framework. This ensures that the problem at hand is focused on in its entirety, there is a discipline to the proceedings and an open-ended problem solving situation does not occur.
    The next section discusses negotiation, the use of social power to achieve a specific end–solving the client's problem as expressed in the value management study.

11.5   NEGOTIATION

The following section draws heavily on the work of Employment Relations (1980), Fisher and Urry (1987) and Kennedy et al. (1984). It is not intended as a thorough exploration of different approaches to negotiation but contrasts and draws together the different perspectives contained within the previously referenced work to provide insights and techniques that can be utilized in value management team situations.

Negotiation is a social process that is concerned with compromise through the use of persuasion. Negotiation takes place because people have no absolute control over events and often exercise their mutual right to differ. It is also about resolving self-interest and conflict through the use of power, either overt or covert, to create movement in attitudes, perceptions or stated positions. Negotiation involves, therefore, understanding the part that perceptions, emotions, motivation and the use of reason have and play in certain types of interpersonal encounter. Negotiation can take place:

• Between individuals;
• Between an individual and a group;
• Between groups.

Schools of thought on negotiation

There are two broad approaches to viewing negotiation. One school of thought, exemplified by Kennedy *et al.* (1984) and to a lesser extent Employment Relations (1980) see negotiation as a process of resolving stated positions. The other school of thought, as exemplified by Fisher and Urry (1987), view negotiation as attempting to resolve interests and not positions. They see positions as clear and explicit but interests are deeper issues, such as, fears, desires and motivations that lie behind positions. Interests can be shared and compatible or conflicting. They also have a strong tendency to be implicit, intangible and inconsistent. However, both schools of thought agree that underlying all types of negotiation are issues of principle and that people and relationships should be separated from the problems to be resolved.

Types of negotiation

Negotiation can be broken down into a number of distinct types. These are:

1. Resolving grievances, perhaps surrounding roles and responsibilities between superior and subordinate;
2. Counselling, for example, staff appraisal schemes or providing advice on aspects of training;
3. Bargaining, for example, agreeing rates, prices or resolving contractual claims;

4. Problem solving, for example, design team meetings or value management exercises.

Resolving grievances and bargaining are undertaken with a greater degree of formality whereas problem solving and counselling involve a lesser degree of formality. Whilst these can be used to categorize different types of negotiation, since it is a social and hence dynamic process the demarcation between types can often become blurred. One determining factor is the presence, in negotiation of mandates or constraints.

Negotiating mandates or constraints

Mandates provide the authority to act and make decisions without reference back to a superior. Mandates can also be viewed as constraints in negotiation since they can either restrict or provide room for manoeuvre. They can sets limits on what can or cannot be agreed and can also provide a useful tactical element in breaking off a negotiation. Mandates can be of two types, fixed, or flexible.

Fixed mandates provide limited room for manoeuvre and will often result in a request to refer back to a higher authority for a decision. They are also likely to push negotiations towards stated positions. Fixed mandates are likely to be present in resolving grievances and bargaining. Flexible mandates, on the other hand, involve limited constraints and will result in a negotiation that is more likely to be fluid and dynamic. Flexible mandates are more likely in problem solving situations or counselling.

Techniques of negotiation

There are a number of useful techniques that can be used in negotiation situations. Some have their place in the preparatory stages of negotiation when strategies are being worked out whereas others are equally important during the process.

Principled negotiation

Fisher and Urry (1987) contend that any negotiation should focus on:

1. Separating the **people** from the **problem**, being hard on the problem and soft on the people by forcing attention away from personal attacks back onto the issues at hand.
2. Targeting **interests** not **positions** since positions can lock people into trains of thought and channel their perceptions in one direction only.

3. Generating copious options before deciding, especially those for mutual gain. This will result in flexible solutions and reduce the likelihood of win/lose outcomes.
4. Judge results against objective criteria, in other words a frame of reference outside the negotiation that both parties can agree on and around which neither had a vested interest.

They suggest these are the basic building blocks for 'principled negotiation', for allowing cases to be judged on their merits.

Role reversal

One of the most commonly referenced techniques in negotiation is to adopt the other side's perspective. It is also the most difficult technique to apply in practice since it may undermine your own viewpoint, may appear threatening or may prove mentally very difficult since it demands a mindset which requires detachment and objectivity. It may also undermine commitment and motivation. However, on the positive side it may increase resolve and assist in preparation since it may facilitate anticipation of possible weaknesses in a case or highlight possible lines of attack from the other side.

Common ground

Common ground can be viewed as the area of agreement between parties to a negotiation. The broader the area of common ground the more likely that agreement can be reached more easily. However, by focusing on common ground when a negotiation has reached an impasse may also generate possibilities for further movement or place areas of disagreement in context.

Aims or objectives

Aims or objectives are the outcomes that are required from a negotiation. They provide a set of criteria against which an agreement can be judged and should be realistic and prioritized. One approach to this is to develop an L-I-M for **both sides** (Kennedy et al. 1984):

LIKE TO GET	Those objectives best considered as 'bonuses'
INTEND TO GET	Those objectives that should be obtained, all things being equal

MUST GET   The bottom line objectives beyond which negoti-
                      ation is broken off if they cannot be obtained.

Fisher and Urry (1987) caution against the use of 'bottom lines'. They argue bottom lines can often be inflexible, unrealistic, too high or set out of context. They prefer to use the idea of a 'trip wire', a notional line beyond which negotiators should consider re-evaluating their objectives in the light of the negotiation that has taken place and after they have had chance to ponder and reflect on the present and likely future outcomes away from the negotiation situation.

A similar technique to L-I-M is the 'Expectation Test' suggested by Employment Relations (1980). The expectation test attempts to structure the premise on which **both sides** enter the negotiation. There are two components to the expectation test:

ISSUES   These would be the issues involved in the negoti-
                      ation. They may well be different or overlap for
                      both sides.
SETTLEMENT POINTS   These are the possible level of concessions that
                      each side would wish to attain from the negotia-
                      tion. Again, there may be similarities as well as
                      differences. These can be considered aims or
                      objectives to be attained.

Each issue has a settlement point. When using the expectation test it is important to adopt both sides' perspective – role reversal. Therefore, for each side there will be a series of issues and associated settlement points. The next stage is to pose the question – **which settlement points will be willingly conceded by both sides**? – A tick should be placed against settlement points willingly conceded by both sides and a cross for those that are unlikely to be conceded easily. The distribution of ticks and crosses for both sides will indicate whether the negotiation is more likely to be orientated towards bargaining (many crosses) or problem solving (many ticks). In addition, those settlement points with ticks against them will provide an indication of possible concessions that both sides may make and the extent of common ground between the parties. The crosses will also indicate issues on which persuasion may be focused in order to convince both sides to change their perspective.

In comparing the two techniques, the L-I-M technique focus attention on prioritizing objectives for both sides. It also indicates where conces- sions can be made. The expectation test focuses attention on both issues and objectives and clearly identified areas of agreement and disagreement. It does, however, lack the prioritizing component of the L-I-M technique. Putting both together provides a very powerful tool for developing a negotiation strategy and in preparing for the process.

Objective criteria

Fisher and Urry (1987) argue strongly that any negotiated agreement should be judged against objective criteria such as market price, precedent, scientifically validated facts, moral standards, tradition, etc. In this way parties to an agreement can judge its merits outwith the vested interests that are present in the negotiation situation.

Tasks

Kennedy *et al.* (1984) contend that in any negotiation situation, as part of the preparation stage, individuals comprising a team should be allocated particular tasks;

1. The leader, who will be the individual with the primary role of presenting a case, questioning, discussing and exploring. This person's principal role is championing a side.
2. The summarizer, who can be considered the 'reserve', the individual that can interject if the leader is running into difficulty but whose primary task is to summarize where the negotiation has got to or the principal points that have emerged. This person's main role is one of clarification.
3. The observer, whose task is to just listen and watch. This is probably the most difficult role to handle since the common tendency is to want to enter a discussion and put your own point of view across. However, the primary focus on looking for verbal and non-verbal cues, sensing changes of mood or negotiation climate/atmosphere and reporting these back to the team during adjournments. The observer is a side's 'sentinel'.

These roles work can work well when there are three or more people in a negotiating team. In a two-person team the summarizer and observer role can be combined. In an individual negotiating role all three have to be combined. The most adept people at combining all three roles are political interviewers and commentators, such as Brian Walden or David and Jonathan Dimbleby.

Questioning

Questioning is a very powerful technique in negotiation. Questioning can shift the balance of power since a person is forced into a response situation, it can provide information and can also clarify. By questioning a person is also taking the initiative.

Carrots and sticks

Carrots are inducements or benefits that can accrue from taking action in a particular direction. Carrots are the positive re-inforcements for action – the reward. Sticks are the unpleasant consequences that can accrue by not taking a particular course or direction. Sticks are the negative reinforcements – the punishments.

Each can be used separately as a technique in negotiation to induce a change. They can also be combined and used simultaneously by highlighting the consequences of both agreement and failure to agree.

Best alternative to a negotiated agreement

Fisher and Urry (1987) suggest the use of a technique which they have dubbed BATNA, the Best Alternative To a Negotiated Agreement. Generally, negotiation takes place in order to produce a result that is better than can be obtained without negotiation. By working out a 'best alternative' a side is forced to consider the options if negotiation fails. Also, the worth of an agreement can be judged by assessing it in terms of the best alternative that could be obtained without negotiating. Developing a best alternative forms part of the preparation phase of negotiation. The process can also be worked out for the other parties to a negotiation.

The best alternative is developed by generating a list of all the future actions that can be taken if no agreement is reached. These are subsequently turned into a series of practical options and the best one selected. This provides the best alternative to a negotiated agreement.

Other issues in negotiation

Adjournments

Adjournments are part of the normal negotiation process. They have advantages and disadvantages. Adjournments can allow a reassessment and re-evaluation of where a negotiation has reached and the objectives obtained or to be attained, it also allows a breathing space for reflection and calm. However, this is a double edged sword for all the parties to a negotiation since the balance of power can shift because all contributors have a chance to re-evaluate the process. The calling of adjournments requires a careful assessment of the 'atmosphere' or 'climate' of negotiation and weighing up the pros and cons of calling a recess.

Deadlocks

Deadlocks can be real or tactical Due to the perceptual distortion that occurs when interpersonal tension is heightened between individuals or groups or within groups, people can 'lock' and become unable to move forward – they have reached a situation of deadlock – a natural process within interpersonal exchanges. An adjournment may assist in the 'unfreezing' process. However, deadlocks can also be a tactical ploy to force concessions (Kennedy *et al.* 1984).

The balance of power in negotiation

This final sub-section places negotiation back within the context of a social process, as outlined earlier. Many of the techniques highlighted above either shift the balance of power between contributors to the negotiation process. For example, questioning, adjournments or deadlocks. Or can be used to assess the relative balance of power between the parties, for example, L-I-M, the expectations test of the best alternative. As Hunt (1986: 69) points out 'power is the deployment of means to achieve ends', in this instance the use of negotiating techniques to move towards agreement. Hunt also contends that power is the ability to affect another person's behaviour without their consent, is rooted in the individual and causes an interpersonal result. However, power is often seen as having manipulative connotations but in Hunt's terms the exercise of power is part and parcel of social processes, especially those present in negotiations.

Negotiation is about reaching compromise through persuasion. The underlying tenet is the use of power in its various forms. Therefore, negotiation is a process of using social power that is ultimately concerned with achieving ends – a mutually acceptable agreement. It is important in any negotiation situation to be sensitive to the subtle shifts in power that can occur as well as anticipating the likely power plays that can be used when preparing for a negotiation or developing a strategy.

The next section looks at chairmanship skills, seen here as the ability and competence to be able to control and guide a team.

11.6   CHAIRMANSHIP SKILLS

A chairman has one of the key influencing roles in any team situation. The role of the 'chair' involves effectively arbitrating, guiding, influencing, controlling and leading a team towards achieving a goal. The key skills of a chairman include (Sydney, Brown and Argyle 1973):

- Dealing with the 'hidden agenda', those issues that are brought to a meeting that on the surface appear unconnected with the task in hand. For example, the ambitions, organizational politics, professional rivalries or prejudices that lurk in the background of any organizational group.
- Recognizing individuals. Each individual comprising a team should be viewed by the chairman as a resource. Failure of an individual to contribute, for whatever reason, may result in lost expertise or a fresh perspective on a problem.
- Questioning and summarizing. The chairman is using one of the key negotiating techniques.
- Providing a sense of direction and common purpose. This is a leadership issue.

The value management team co-ordinator is acting in the role of chairman of a team. The skills outlined above are particularly important in the VMTC role. The final section draws together the issues raised in this chapter within the context of value management.

11.7   SUMMARY

Value management is a multi-disciplinary, team-orientated approach to problem solving. To be effective the VMTC, as chairman and facilitator, has to harness the creativity of the team and direct it towards solving the client's problem through functional analysis. In order to do this the VMTC has to understand the various stages through which a value management team, as a temporary work group develops and progresses. This involves the VMTC in skills of group diagnostics and chairmanship as well as negotiation. Therefore, the VMTC has to be adept at using social power in conjunction with problem-solving skills to steer the value management team towards arriving at a solution or solutions to the client problem, usually expressed in functional language. The VMTC has to have at his or her disposal tools and techniques that are relevant to each stage of group development and to the problem solving process. The VMTC also has to be adept at sensing and managing changes in the 'atmosphere' or 'climate' of the team, to shifts in power between individuals comprising the team and within the group itself. There may also be changes in the power structure within which the value management team operates in its external environmental context. The next chapter presents a proposed UK method for project economics.

CHAPTER 12

A PROPOSED UK METHODOLOGY

12.1 VALUE MANAGEMENT

Value management is:

> a service which maximizes the functional value of a project by managing its development from concept to occupancy through the audit (examination) of all decisions against a value system determined by the client.

This service is achieved through the Job Plan which remains basically as that proposed by Miles in 1961 and as illustrated in Table 2.2. At the core of a value management service is the identification of **function**, which is defined as:

> an activity for which a thing is specifically designed, used, or for which something exists.

and **value** which is defined as:

> a measure expressed in money, effort, exchange or on a comparative scale which reflects the desire to obtain or retain an item, service or ideal.

Function and its associated value can only be determined by reference to the client or end-user.

12.2 THE CONSTITUENTS OF A PROJECT ECONOMIC SERVICE

There are three basic components to a project economics service illustrated in Figure 12.1, namely, value management, structured cost studies and traditional cost consultancy provided by the quantity surveying profession.

158

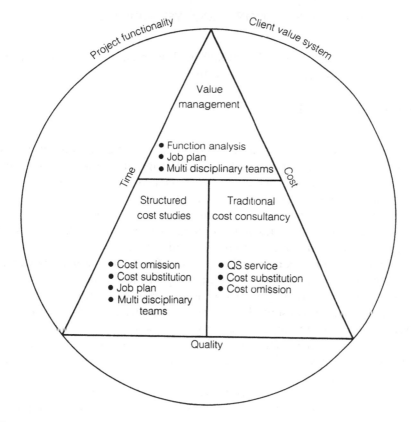

Figure 12.1   Project economics.

A value management service is defined as one that involves:

- functional analysis;
- life-cycle costing;
- operating in multi-disciplinary work groups using the job plan and creativity techniques;
- establishing comparative cost in relation to function and hence overtly concerned with issues of value.

A structured cost study is defined as one that involves:

- operating in multi-disciplinary work groups using the job plan and creativity techniques;
- cost reduction, cost substitution and/or specification changes.

Cost consultancy provided by the quantity surveying profession tends to result in cost reduction or cost substitution.

12.3   A PROPOSED FUNCTIONAL ANALYSIS METHOD

There are four methods of functional analysis relating to the four levels illustrated later in Figure 12.3. These are:

Level 1 – Project task

As an alternative to the traditional method of project procurement, where an architect may be approached first, it is suggested that a client's first approach should be to a value manager (facilitator) who with representatives of the client organization will carry out a value audit and enable the client to determine whether a built solution is the best answer to the problem. Additionally, if this is so, criteria that may impact the choice of the best procurement route can also be made explicit.

A building accommodates activities. Therefore a functional analysis of the building concept should concentrate upon the client's project task – the reason or reasons why a project is required in the first place. In all cases this can be represented by a process flowchart. Even in the most simple case, for example, a house, the occupants use the spaces for an activity and often these activities follow a logical order, ie., putting on a coat and a pair of shoes before leaving. These activities must be satisfied by space provision.

In the analysis of the organizational processes it is necessary to flow-chart the means of carrying out the activity within an organizational framework. This follows the established concept that value management is an audit process. The proposed procedure is as follows.

(a)   Determine the prime function

It is important to accurately determine the prime function through a clear understanding of the client's objectives. For example, an entertainment company has realized the potential for a small quality cinema in a particular area. The prime function therefore, must relate to their objective which is to 'show modern popular films in a quality environment for profit'. This allows brainstorming to be directed towards the subject. The prime function of 'to make money' is incorrect as this would allow brainstorming to consider all means of making money. However, the client, in this case, is in the business of entertainment and the prime function has to be rooted in the context of the client's activity.

It should be noted that the primary function was realized through the identification of a technical solution, namely, 'a small quality cinema'. Functional analysis of the technical solution provides a clearer statement of the objective to 'show modern popular films in a quality environment for profit'. This functional definition does not specify a built solution and other options may be available and subsequently explored.

(b)   Identify the process

The next step is to identify the process to be carried out within the facility. For example, the process to enable people to 'see films in a quality environment' is as follows:

(i)   advertise film;
(ii)   receive people;
(iii)   receive payment;
(iv)   direct people to seat;
(v)   show film;
(vi)   signify end;
(vii)   direct people to exit.

The identification of the process has to be agreed with the client. It has to be accepted that this process is described from a basis of a perceived technical solution.

(c)   Ask the question HOW?

The HOW/WHY logic of a technical FAST diagram is not applicable to the process.

However, for each stage in the process the question how? should be asked. For example, 'receive people' may be sub-divided by asking the question how? into 'welcome', 're-assure', 'make comfortable'; all these are functions, any or all of which may be the subject of brainstorming to achieve a technical solution.

On the other hand asking the question how? of the function 'direct people to seat' may throw up the answers 'physically lead them', 'sign-posts', etc.; all technical solutions for which no function other than 'direct people to seat' may be found, proving that this function cannot be divided into sub-functions.

Incidentally, the technical solutions generated at this stage should be recorded so that they may be placed on the table at the commencement of brainstorming. These are incorporated into the results of brainstorming and judged during the judgement phase. This is a 'participant mind dump', which clears the mind of technical solutions which have to be remembered later and also standard solutions that may hinder exploration of more novel ideas.

(d)   Commence brainstorming

Brainstorming now commences on the derived functions. The normal rules of brainstorming apply and at this stage there should be no bar to the number, type or relevance of the suggestions made. They should all be recorded.

(e)   Judgement and brief

All ideas on a common theme should be grouped together. These may be randomly spread through the list of ideas and therefore should be grouped and collated under a heading. If the client representative does not wish to consider ideas on a particular theme then these should be deleted from further study. Those ideas which are judged viable are carried forward to a development stage where their space and arrangement implications are investigated. From the data derived, the basis of the brief which reflects the process to be accommodated within the building, is committed to paper.

An important conclusion at the end of level 1 is whether or not a built solution is anticipated. Either way a detailed record of client requirements is to hand.

Level 2 – Spaces

Having determined that a building is the most promising solution to the problem the VM process moves to level 2 which conceives of a building as an environmentally controlled space. At this level the aim is to determine the functional requirements of the space which can be carried out in three ways:

1. The analysis of space as it is seen to be required during the development of the brief through the medium of value management. (Pre-brief);
2. The analysis of space as it is seen to be required by the client's brief. (Concept);
3. The analysis of space as it is represented on the architect's sketch drawings. (Schematic).

The requirements of space will be of size, environment, adjacency, access, etc. Following the analysis a study is made of those spaces which have the same or similar characteristics. This will allow combining, zoning and rationalization. For example, spaces which have a similar specific heating and ventilating requirement could be together. This is part of the rationale in case study 3, the hospital, in Chapter 4. This may, for example, allow the cost of keeping them separate or the implications for organizational functioning to be realized. Also, spaces which have an identical function may be shared or combined, etc.

Level 3 – Elements

An element has a function but does not perform or contribute directly to the process being carried out within the building. For this reason it could be assumed that a value management exercise can be undertaken in the absence of the client although some input may be required at some stage to validate the aesthetic and maintenance implications.

Elements perform functions which are not generally project-related. It should, therefore, only be necessary to carry out a functional analysis of an element once. Functions having been realized for an element in one building are then relevant to the majority of buildings. Therefore a level 3 functional analysis could be applied on an industry-wide and not a project specific basis.

For example, the function of a foundation is always to distribute load, the function of a glass window is generally to permit one or more of the following:

Primary Functions

- Transmit light;
- Control ventilation;
- Prevent inward access;
- Retain heat;
- Insulate from external noise;
- Create view (one way or both ways);
- Improve means of escape.

Secondary Functions

- Amplify solar heat gain;
- Increase distracting sunlight;
- Establish the generation of cold radiation.

The number of functions appropriate to a specific project may be limited by context. For example, consider a prestigious, multi-storey office building for an owner-occupier to be constructed close to a motorway exit and within the sound footprint of a neighbouring airport. The building will be sealed against external noise and will be air-conditioned. The window will therefore not be required for the function 'control ventilation'. In fact, it is probable that the function 'transmit light' is also not required as the building, being air-conditioned, is likely to be deep plan. From the list of primary functions, 'prevent inward access', 'retain heat', 'insulate from external noise', 'improve means of escape' could either be better provided by the external wall construction or are not relevant in this building.

The 'secondary functions' by definition are not required and are only present because of the choice of a sheet of transparent material for the purposes outlined. It is interesting that many designers put more effort into solving the secondary function problems rather than analyzing the primary functions. In this example some designers would go to great lengths to reduce solar heat gain by the use of special glass, electrically controlled external blinds, internal blinds, etc., all of which impair the function create view', which in the above example is probably the only function left to be provided.

A better approach is to consider the function 'create view' and brainstorm the ways in which this may be done. TV cameras and large flat screen televisions may be an alternative. This would allow considerable design freedom both internally and externally.

In carrying out such an exercise it is important that required functions, not on the standard list due to the uniqueness of the project, are not left out.

Level 4 – Components

Similar to elements above, components have a function but do not perform or contribute directly to the process being carried out within the building. For this reason it can be assumed that a value management exercise could be undertaken in the absence of the client although some input may be required at some stage to validate the aesthetic and maintenance implications. It will be necessary, however, to have an input from the supplier particularly in those situations where a relatively novel solution has been found to an elemental function.

Components used in building have been subjected to a process prior to their arrival on site. All or parts of this process may be analysed to determine function and having determined function a brainstorming exercise may yield alternative technical solutions to the process. Understanding the manufacturing process and the functional requirements of the component may lead to alternative design decisions particularly when dealing with components which are project specific. For example, curtain walling, precast-concrete components, windows and doors. Consider a precast-concrete cladding panel, it is subjected to a process of manufacture, transport, lifting and fixing in position. Each one of these functions has a number of technical solutions. The manufacture itself is a process comprising a number of discrete functions. This level has the closest affinity to manufacturing orientated value management and the tools and techniques of, for example, technical FAST are applicable here.

12.4   ALTERNATIVE STUDY APPROACHES

This section details nine different study approaches, excluding normal quantity surveying services, that could be chosen as part of a project economics service. The choice of approach will depend on the client problem to be addressed and the stage of the project life-cycle that is to be targeted for study.

There are five basic approaches to undertaking a value management study illustrated in Figure 12.2. These are termed:

(a)   *VM1,* where the personnel involved are a study facilitator and client representatives who have an input to the definition of the project.

The primary task of this team is to identify: the project task, client needs and wants, stated explicitly in functional terms. They will primarily be concerned with level 1 but may address level 2 issues. The aim of this audit is to ensure that the client requires a built solution.

(b)   *VM2,* where the personnel are a study facilitator and client representatives together with an independent audit design team. This design team may not be involved with the project design. At this point the decision to build has been taken but the procurement route is still open.

The primary task of this team is to address level 1 and level 2 issues. This allows the procurement route to be the subject of investigation when strategic project issues are explored with the client. It may also lead to faster procurement since the project may be appraised functionally and technically prior to the brief being detailed for design.

(c)   *VM3,* where the personnel are a study facilitator and client representatives together with the project design team. At this point the decision to build has been taken and the appointment of a project team would

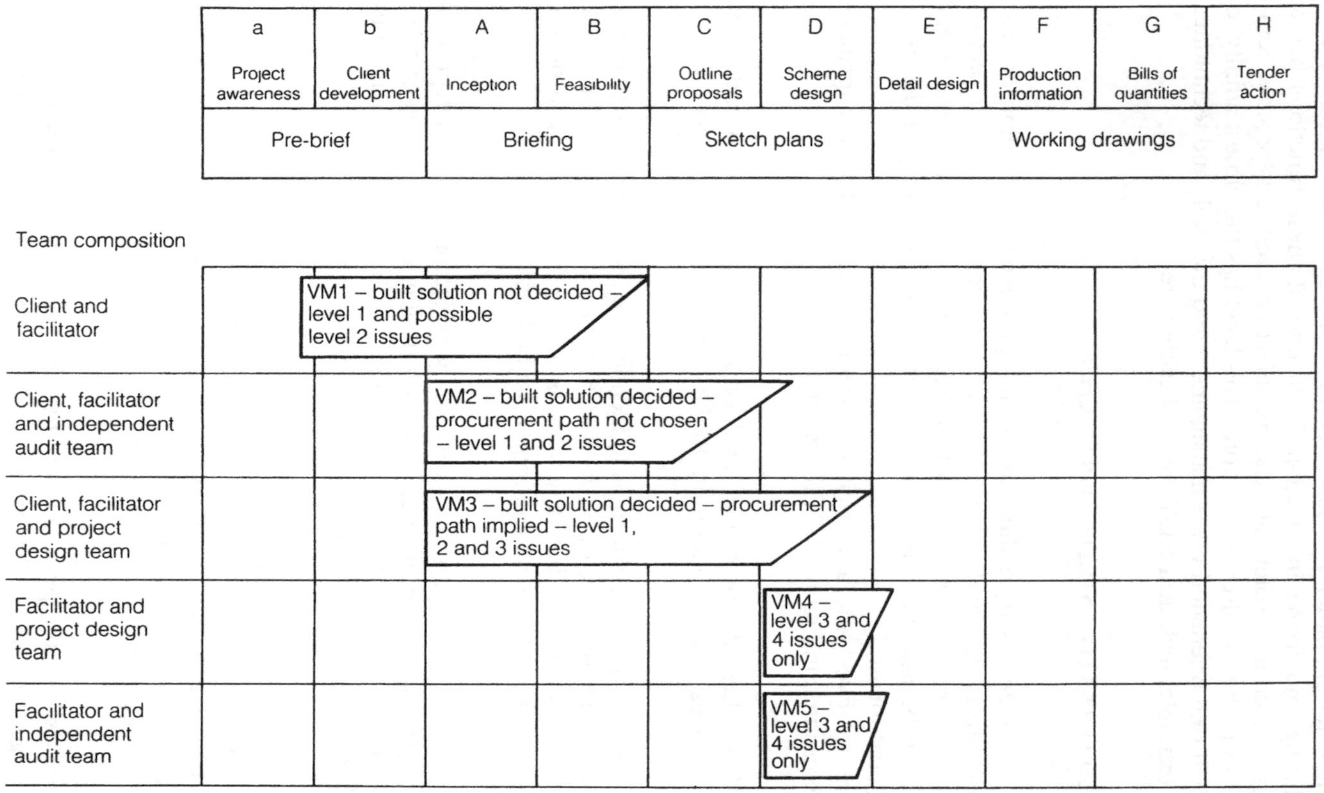

a	b	A	B	C	D	E	F	G	H
Project awareness	Client development	Inception	Feasibility	Outline proposals	Scheme design	Detail design	Production information	Bills of quantities	Tender action
Pre-brief	Briefing			Sketch plans		Working drawings			

Team composition

Client and facilitator	VM1 – built solution not decided – level 1 and possible level 2 issues								
Client, facilitator and independent audit team			VM2 – built solution decided – procurement path not chosen – level 1 and 2 issues						
Client, facilitator and project design team			VM3 – built solution decided – procurement path implied – level 1, 2 and 3 issues						
Facilitator and project design team					VM4 – level 3 and 4 issues only				
Facilitator and independent audit team					VM5 – level 3 and 4 issues only				

Figure 12.2 The five basic approaches to value management. The area denotes the most likely focus in time for the study.

indicate a traditional procurement route but with the options for construction management or management contracting still open. This is seen to be non-conflict orientated and with increased possibilities for implementation. The exercise can be used for briefing the design team and maximizes project knowledge. This exercise is seen to be inexpensive since the design team are not likely to charge an additional fee for this briefing exercise. This has a close affinity to the charette.

(d)   VM4, where the personnel involved are a study facilitator together with the project design team. The project will have reached sketch design. While the team may address level 2 issues, in the absence of a client representative they are more likely to be concerned with levels 3 and 4. The absence of the client would indicate that the client's value system is embedded in the design to date. This is non-conflict orientated but can increase the design period particularly where reference back to the client is required.

(e)   VM5, where the personnel involved are a study facilitator together with an independent audit design team. This is the North American value engineering situation where functional analysis is used. The project will have reached sketch design. While the team may address level 2 issues, in the absence of a client representative they are more likely to be concerned with levels 3 and 4. The absence of the client would indicate that the client's value system is embedded in the design to date. This provides objectivity and the opportunity for increased technology transfer. However, there is an increased risk of adversarial relationships developing and can also increase the design period.

There are four basic approaches to undertaking a structured cost study as illustrated in Figure 12.3. These are termed:

(a)   SCS1, where the personnel are a study facilitator and client representatives together with an independent audit design team. This design team will not be involved with the project design. The project will have reached sketch design stage. This provides objectivity and the opportunity for technology transfer. However, adversarial relationships with the existing design team are a possibility.

The primary task of this team is to address level 3 and level 4 issues. However, there may be small changes to level 2 issues.

(b)   SCS2, where the personnel are a study facilitator and client representatives together with the project design team. The project will have reached sketch design. This is non-conflict orientated but the incentive for change is limited.

The primary task of this team is to address level 3 and level 4 issues. However, there may be small changes to level 2 issues.

(c)   SCS3, where the personnel involved are a study facilitator together with the project design team. The project will have reached sketch design and team will address level 3 and 4 issues. This has close similarities to the North American concurrent workshop. The absence of the client would indicate that the client's value system is embedded in the design to date. This is non-conflict orientated but the incentive for change is limited.

(d)   SCS4, where the personnel involved are a study facilitator together with an independent audit design team. This is the widely adopted North American value engineering situation.

The project will have reached sketch design and while the team may address level 2 issues, in the absence of a client representative are more likely to be concerned with levels 3 and 4. The absence of the client would indicate that the client's value system is embedded in the design to date. This is adversarial but is objective and with the opportunity for technology transfer. However, the implementation rate of potential savings tend to be low and there is a danger that quality can be compromised.

A quantity surveying cost study is most likely to address level 3 and 4 issues as illustrated in Figure 12.3 following the completion of the sketch design and the cost plan.

12.5   AN EXAMPLE OF A UK CASE STUDY

Steven Male acted recently in the role of a facilitator on a £200 million UK based mass-transit system that was well into sketch design. A multi-disciplinary team was used, with both a client representative and operational manager in the group. The primary focus of the study was similar to type SCS1 as described above. The study was conducted over a number of weeks – a concurrent workshop – just after sketch design stage and in some instances well into production drawings.

The study did not use functional analysis *per se* but utilized the interactive nature of a multi-disciplinary team to analyse the project using their collective problem solving skills. A number of different types of potential savings were identified which can be classified as:

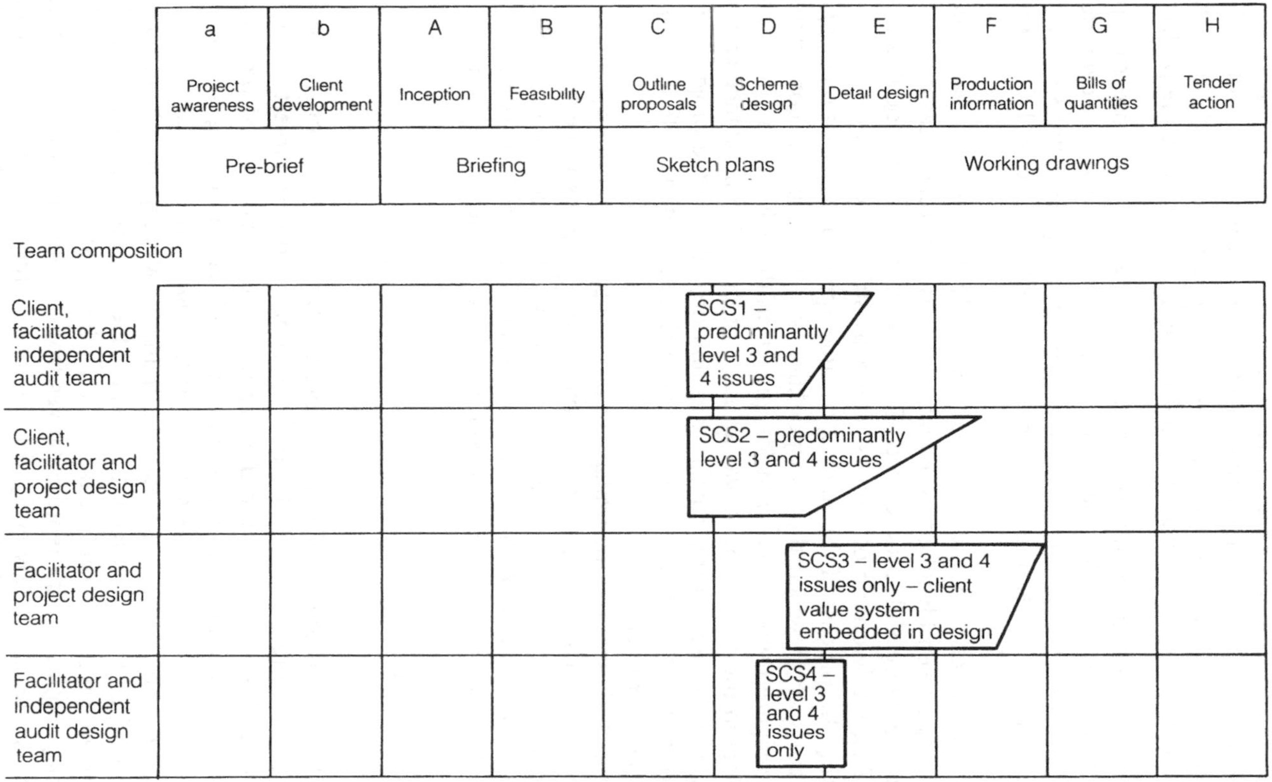

a	b	A	B	C	D	E	F	G	H
Project awareness	Client development	Inception	Feasibility	Outline proposals	Scheme design	Detail design	Production information	Bills of quantities	Tender action
Pre-brief		Briefing		Sketch plans		Working drawings			

Team composition

Client, facilitator and independent audit team				SCS1 – predominantly level 3 and 4 issues					
Client, facilitator and project design team				SCS2 – predominantly level 3 and 4 issues					
Facilitator and project design team						SCS3 – level 3 and 4 issues only – client value system embedded in design			
Facilitator and independent audit design team					SCS4 – level 3 and 4 issues only				

Figure 12.3 The four basic approaches to cost management. The area denotes the most likely focus in time for the study.

1. Typical cost reduction/substitution/omission items.
2. Cost savings obtained from re-configuring the use of space allocations.
3. Cost savings due to modifications to existing ideas in the design.
4. Cost savings due to an alteration to the concept that also involved (1) and (2) above but also potentially adding value to the client in the longer term. In this instance something new was added to the project.

Even although the study is designated as being very close to type SCS1 above the study identified issues at levels 1, 2 and 4. However, due to the advanced stage of the project there were potentially significant problems in implementing those changes identified as level 1 (concept) issues.

At the completion of the study approximately 8% of the estimated project cost had been identified as potential savings and some areas of potential savings were still to be costed. These savings had been identified as part of the study on top of a series of already stringent cost reviews.

During the process of the study it became apparent also that one single item, the way in which the transportation system itself was constructed and configured would have a major impact on the value of the project to the client in the long term. It was an issue of strategic importance and an emerging statement of the primary function/objective of the project highlighted the fact that value/cost issues were closely related to life-cycle and competitive criteria.

As the study evolved the primary task of the project became apparent to the facilitator. The statement that captured the project's primary task involved aspects of the client's corporate strategy that embodied this project. The life-cycle issues of the project in relation to substitute competition from other transport systems also emerged as a significant question.

12.6 CONCLUSIONS

This chapter has identified clear distinctions between levels of functional analysis on a project. Levels 1 and 2 are concerned with project task and concept and are embedded in the organizational structure and processes of the client and are intimately connected with understanding the client's value system. Levels 3 and 4 relate to the built structure, the technical solution to the client's functional requirements. Moving from level 1 to 4 increases the extent to which the client value system is embedded in the design. At level 1 – project task – the client value system has to be explicitly explored with representatives from the key activity areas of the client. At level 4, however, the client's value system is implicitly incorporated into the design whether or not this was made explicit in the brief.

This chapter has also outlined the types of study that can be utilized at different stages of the project life-cycle, indicating the personnel and possible advantages and disadvantages for each.

CHAPTER 13 _____

KEY ISSUES FOR PRACTICE

_____

13.1 INTRODUCTION

In this final chapter the term value engineering has been used to identify North American practice, whereas value management is used to denote a broader approach to a management function concerned with value, and reflects the authors' view of an appropriate term for use in the UK.

In the US the consequences of heavy public sector patronage through requirements of accountability and the absence of a QS input has resulted in a standard approach to value engineering. This is a 40 hour independent, multi-disciplinary workshop at 35% design which seeks to demonstrate value for money. Research evidence suggests that this is a combination of cost reduction or cost substitution or changes to project concept. In a UK context there is a similarity between a part of this approach and cost planning.

There is an apparent paradox in the US view of value engineering in that US authors state that VE included early in the design process offers maximum benefits. The majority of VE practice occurs at 35%, or sketch design, where a detailed specification and a full cost estimate exists and from which it is easy to identify cost reduction or substitution items, and hence demonstrate quantifiable savings.

The research in the US demonstrated that there is a diversity of practice not reflected in the text books and the core of value engineering revolved around the use of functional analysis, used as a tool to varying degrees by practitioners. The proposed UK methodology in Chapter 12 has attempted to extract the strengths from this diversity of practice.

13.2 VM PRACTICE IN THE UK

Value management re-defined

VM is a proactive, creative, problem-solving service. It involves using a structured, multi-disciplinary team-orientated approach to make explicit the client's value system using function analysis to expose the relationship between time, cost and quality. Strategic and tactical decisions are audited against the client's value system at targeted stages through the development of a project or the life of a facility.

Within the above definition strategic decisions are seen as those made in the pre-brief, brief or concept stages of a project. Tactical decisions can be seen effectively as those occurring after 35% design.

Within value management, four levels of functional analysis have been identified: levels 1 and 2 – project task and spaces require a systems view of the project, thinking in terms of concepts and wholes. Levels 3 and 4 – elements and components define sub-systems and are more detailed in their requirements for thought processes. Functional analysis should be looked at in terms of an exploration of project characteristics as defined by the value management team and client and evaluated in life-cycle cost terms. It is important in this exploration of function to explicitly determine client needs and wants.

VM team characteristics and the role of the VMTC

The choice of VM team has to reflect project characteristics and client input into the team is essential up to 35% design. Functional analysis re-orientates the perceptions of a team about the task of a project through group dynamics, hence the requirement for client involvement. However, even excluding functional analysis, the use of brainstorming through the dynamics of a multi-disciplinary team are useful in auditing a project – this has been termed **cost management** in this book as opposed to **value management**.

The roles of VMTC and the cost consultant should be kept separate. Large multi-disciplinary practices appear better positioned to offer a diversity of VM services and standardize operating procedures. In comparison, smaller and/or monoprofessional practices will require a network of consultants who can be pulled together to suit the project.

A VM exercise places considerable demands on the facilitator (VMTC) for interpersonal skills, involving negotiation, persuasion, chairmanship, the use of social power and the understanding of group dynamics.

A VM service offering primarily the involvement of a VMTC or facilitator and support staff using the existing design team would appear to be more appropriate in a UK context. A high probability of the implementation of VM recommendations and reduced levels of conflict can be seen as the main advantages. The major disadvantage could be a lack of objectivity by the existing design team about past decisions. This latter point is seen as the main advantage of the independent VE team used in North America.

The client's role in value management

The research demonstrated that the influence of top management on the implementation process within the client organization is essential. However, the effects of organizational structure on the locus of power for implementation, namely decentralized versus centralized decision making, cannot be ignored. Also, the target group responsible for implementation is a key factor. Regular procurers of VM services, are more likely to evolve standardized operating procedures. Regular procurers in North America would normally expect savings in the order of 8 to 25% with an outlying range between 3 – 30%. This contrasts with a probable saving of between 5 – 10% using traditional QS techniques in the UK.

VM and the management of projects

Value management cannot be divorced from the management of projects and provides a mechanism for integration to improve communication and information flows. This allows the exploration and interrelationship between time, cost, quality and function and any trade off between these.

The value management of major projects, which involves issues of size, complexity, schedule urgency or high demands in terms of existing resources and know-how, provides the opportunity to anticipate and explore potential problems and solutions associated with these issues. However, this will require the appropriate choice of VM team, the correct timing of the study and sufficient time being devoted to project analysis and solution building.

Where a client engages a value manager on major projects and where an independent VM team is used, multiple studies using the same team on the same project provides the opportunity for retention of project specific knowledge and overcomes one of the disadvantages of using an independent audit.

The research evidence has questioned the validity of the 40 hour single shot workshops at 35% design regardless of project size and complexity. Due to the requirements of information assimilation, group dynamics and time limitations, 40 hours was suggested as being satisfactory on small to average sized projects whereas a range of between 40 to 100 hours was given for complex projects especially those with high levels of M&E. One organization appeared to optimize on a 56 hour study broken down into periods of 3 days, 2 days and 2 days of team activity spread across 3 weeks. This would appear to minimize the disruption to work routines, allow information assimilation and subconscious problem solving processes to work.

Terms of engagement & liability

Where a VM exercise is likely to be undertaken on a project the terms of engagement of the project design team should reflect this. The design liability surrounding VM recommendations could be a contentious issue where an independent VM team is used.

Project economics

This book has also advocated a service – termed **project economics** – which encompasses the integration of value management and cost management procedures. Part Four of this book provides a method and set of tools and techniques for use in a UK context as part of this project economics service.

REFERENCES

Adair, J. (1987) *Effective Team Building*. Pan, London

Adams, J. L. (1987) *Conceptual Blockbusting: A guide to better ideas*, 3rd edn, Penguin, Harmondsworth.

Borjeson, L. (1976) *Management of Project Work*. The Swedish Agency for Administrative Development, Satskontoret, Gotab, Stockholm.

Broadbent, G. (1973) *Design in Architecture*, Wiley, Chichester.

Brun, G. (1989) Marketing and value analysis, Proceedings of 1992: *Value Management, the Key to Success in European Business*. 1st European Conference on Value Management, October 16th–17th, Milan, Italy, 415–20.

Burt, M. E. (1975) *A Survey of Quality and Value in Building*, BRE.

Chartered Institute of Building. (1982) *Project Management in Building*. Report.

Checkland, P. B. (1981) Science and the systems movement, in *Systems Behaviour,* 3rd edn. Open Systems Group, Harper and Row, London, 288–314.

Churns, A. B., and Bryant, D. T. (1984) Studying the client's role in construction management. *Construction Management and Economics*, **2** (1), 177–84.

Cross, N. (1989) *Engineering Design Methods*. Wiley, Chichester.

Crum, L. W. (1971) *Value Engineering: The organised search for value*. Longman, London.

de Bono, E. (1970) *Lateral Thinking*. Penguin, Harmondsworth.

de Bono, E. (1982) *de Bono's Thinking Course*. Holland, London.;

de Greene, K. B. (1970, 1981) Systems and psychology, in *Systems Behaviour*, 3rd edn., Open Systems Group, Harper and Row, London 83–118.

Dell' Isola, A. J. (1974) *Value Engineering in the Construction Industry*. Construction Publishing Co., New York.

Dell' Isola, A. J. and Kirk, S. J. (1981) *Life-cycle Costing for Design Professionals*. McGraw-Hill, New York.

Dell' Isola, A. J. and Kirk, S. J. (1983), *Life cycle cost data*, McGraw-Hill, New York.

176

Employment Relations (1980) *Introduction to Negotiations*. Compass House, Cambridge.

Fallon, C. (1971) *Value analysis to improve productivity,* Wiley, New York.

Fisher, R. and Urry, W. (1987) *Getting to Yes.* Arrow, London.

Flanagan, R. and Norman, G. (1983) *Life-cycle Costing for Construction.* RICS, Surveyors Publications.

Flanagan, R., Norman, G., Meadows, J. and Robinson, G. (1989) Life-cycle Costing – Theory and Practice. BSP Professional Books,

Gage, W. L. (1967) *Value analysis.* McGraw-Hill, New York.

Hellriegel, D., Slocum, J. W. and Woodman, R. W. (1986) *Organisational Behaviour,* 4th ed. West Publishing Co., St. Paul.

Honey, P. and Mumford, A. (1982) *The Manual of Learning Styles,* Honey, Maidenhead.

Hunt, J. W. (1972) *The Restless Organisation,* Wiley, Chichester.

Hunt J. W. (1986) *Managing People at Work,* 2nd ed. McGraw-Hill, London.

Janin, L. F. (1989) *Functional specifications.* Proceedings of International Conference–Society of American Value Engineers, June 11–14, Indianapolis, Indiana, USA, **XXIV**, 112–17.

Janis, I. L. (1972) *Victims of Groupthink: A psychological study of foreign policy decisions and fiascos.* Houghton Mifflin, Boston.

Janis, I. L. and Mann, L. (1977) *Decision Making: A psychological analysis of conflict, choice and commitment.* Free Press, New York.

Jenkins, G. M. (1969, 1981) The systems approach, in *Systems Behaviour* 3rd ed, Open Systems Group. Harper and Row, London, 288 – 314.

Jones, J. C. (1981) *Design Methods.* Wiley, Chichester.

Kast, F. E. and Rosenzweig, J. E. (1972, 1976) The modern view: a systems approach, in *Systems Behaviour,* 2nd ed (eds J. Beishen and G. Peters), Open Systems Group, Harper and Row, London.

Kelly, J. R. and Male, S. P. (1987) A Study of Value Engineering and Quantity Surveying Practice, Research Report for QS Division, Royal Institution of Chartered Surveyors, Heriot-Watt University, Edinburgh.

Kelly, J. R. and Male, S. P. (1988) *A Study of Value Management and Quantity Surveying Practice,* Occasional Paper, Surveyors Publications, London.

Kelly, J. R. and Male, S. P. (1991) *The Practice of Value Management: Enhancing value or cutting costs?,* RICS, London.

Kennedy, G., Benson, J. and McMillan, J. (1984) *Managing Negotiations,* 2nd edn. Hutchinson, London.

Litaudon, M. (1989) Value engineering in big equipment projects: new concepts to control conception in time. Proceedings of *1992: Value Management, the Key to Success in European Business,* 1st European Conference on Value Management, October 16th–17th, Milan, Italy, 463–74.

MAC (1985) Competition and the Chartered Surveyor: Changing client demand for the services of the chartered surveyor. Report by Management Analysis Centre for the Royal Institute of Chartered Surveyors, HMSO, London.

Macedo, M. C., Dobrow, P. V. and O'Rourke, J. J. (1978) *Value Management for Construction*. Wiley Interscience, New York.

Male, S. P. (1991) *Competitive Advantage in Construction*. Butterworth-Heinemann, Oxford.

Male, S. P. and Kelly, J. R. (1989) Organisational responses of public sector clients in Canada to the implementation of value management: lessons for the UK construction industry. *Construction Management and Economics*, **7** (3), Autumn 1989, 203–16.

Markus, T. A. (1967) The role of building performance measurement and appraisal in design. *Architects Journal*, 20 December, p. 1567.

Marshall, H. E. (1988) *Techniques for treating uncertainty and risk in the economic evaluation of building investment*, US Department of Commerce, NIST special publication 757.

Miles, L. D. (1972) *Techniques for Value Analysis and Engineering*, 2nd ed. McGraw-Hill, New York.

Morris, P. W. G. and Hough, G. H. (1987) *The Anatomy of Major Projects: A study of the reality of project management*. Wiley, Chichester.

NEDO (1987) *Faster Building for Commerce*. National Economic Development Office, Millbank Tower, London.

O'Brien, J. (1976) *Value Analysis in Design and Construction*. McGraw-Hill, New York.

Palmer, A. (1992) A comparison. of US value engineering with British cost control procedures. Seminar: *Value and the Client*, Royal Institution of Chartered Surveyors, London, January 29.

Parker, D. E. (1985) *Value Engineering Theory*. Lawrence D. Miles Foundation, Washington D. C.

PRS (1987) The Architect in a Competitive Market. Report by Property Research Services for the Cities of London and Westminster Society of Architects and London region of RIBA.

Porter, M. E. (1985) *Competitive Advantage*. Free Press, New York.

Reve, T. (1990) The firm as a nexus of internal and external contracts. eds, Aoki M., Gustafsson B. and Williamson O. E., T*he Firm as a Nexus of Treaties*, Sage Publications.

RISC (1991) *Quantity surveying 2000: the future role of the chartered quantity surveyor*, Royal Institution of Chartered Surveyors, London.

Ruegg, R. and Marshall, H. (1990) *Building economics; theory and pratice*, Van Nostrand Reinhold, New York.

Sayles, L. R. and Chandler, M. K. (1971) *Managing Large Systems: Organisations for the future*. Harper and Row, New York.

Schein, E. H. (1980) *Organisational Psychology*, 3rd ed. Prentice-Hall, Englewood Clis, New Jersey.

Snodgrass, T. J. and Kasi M. (1986) Function analysis. University of Wisconsin, USA.

Sydney, E., Brown, M. and Argyle, M. (1973) *Skills with People*. Hutchinson, London.

UCERG (1975) *The Mechanism of response to effective demand: the supply of construction services in the West Midlands Building Economics Research Unit*, University College Environmental Research Group, London.

US General Accounting Office (1978) *Computer-aided building design*, US Department of Commerce, Washington DC.

Vickers, G. (1980, 1981) Some implications of systems thinking in *Systems Behaviour*, 3rd ed, Open Systems Group. Harper and Row, London, 19–25.

von Bertalanffy, L. (1962, 1981) General systems theory – a critical review, in *Systems Behaviour*, 3rd ed, Open Systems Group. Harper and Row, London 59–79.

Walker, A. (1989) P*roject Management in Construction*, Blackwell Scientific Publications, Oxford.

Wilson, B. (1990) *Systems: Concepts, methodologies, and applications*: 2nd ed. Wiley, Chichester.

Winch, G. (1987) The construction firm and the construction process: the allocation of resources to the construction project, in *Managing projects worldwide* (eds Lansley, P. R. and Harlow, P. A.), **VII** 967–75, E&FN Spon, London.

Zimmerman, L. W. and Hart, G. D. (1982) *Value Engineering: A practical approach for owners, designers and constructors*. Van Nostrand Reinhold, New York.

INDEX